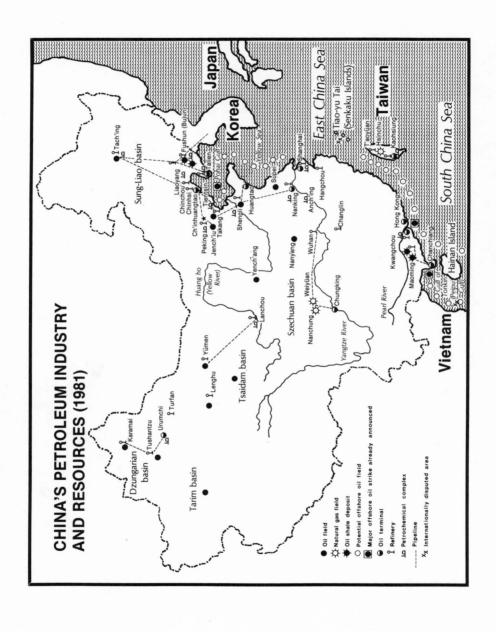

CHINA'S PETROLEUM INDUSTRY
AND RESOURCES (1981)

CHINESE PETROLEUM

an annotated bibliography

A
Reference
Publication
in
Asian
Studies

Frank Joseph Shulman
Editor

CHINESE PETROLEUM

an annotated bibliography

RAYMOND CHANG

中國石油

G.K. HALL & CO.

70 LINCOLN STREET, BOSTON, MASS.

Library of Congress Cataloging in Publication Data

Chang, Raymond J.
    Chinese petroleum.

    Includes indexes.
    1. Petroleum industry and trade — China — Bibliography.
I. Title.
Z6972.C44   [HD9576.C52]   016.3382′7282′0951   82-2930
ISBN 0-8161-8333-3   AACR2

This publication is printed on permanent/durable acid-free paper
MANUFACTURED IN THE UNITED STATES OF AMERICA

Contents

The Author

Raymond Chang holds a B.A. in Foreign Language from Taiwan University, an M.A. in History from New York University, and an M.L.S. from Long Island University. He also earned an Advanced Professional Certificate from the Graduate School of Business Administration at New York University. Currently he is an associate professor and bibliographer at the Library Department of Bernard Baruch College of the City University of New York, specializing in energy, finance, and international trade. His writing includes articles on ancient Chinese printing published by the Christian Science Monitor and the Journal of Library History, and Petroleum Industry in China: A Bibliography, published by the Council of the Planning Librarians.

Introduction

China, once considered an oil-scarce country, is not only self-sufficient in petroleum and its products, but is also on the threshold of becoming one of the world's leading oil producers. The growth of the petroleum industry in China since 1950 has been phenomenal. Crude oil production rose from less than half a million tons in 1952 to 106 million tons (2.1 million b/d) in 1979--more than a 200-fold increase in 27 years. The development of China's petrochemical industry is equally rapid and impressive. This remarkable feat has been achieved mainly in the spirit of self-reliance and independence, with minimal foreign assistance. The contribution of the oil industry to China's economy has been so substantial and politically so important that today the Taching oil field, China's leading oil base, is hailed as an industrial model for all of China's industrial enterprises.

With the downfall of the "Gang of Four" and the establishment of a more pragmatically oriented government, China has embarked upon an ambitious program of modernization in four areas (agriculture, industry, science and technology, and national defense), with the aim of transforming an economically underdeveloped country into a modern and advanced industrial power by the turn of the century. The role that the petroleum industry will play in realizing China's national goal will be pivotal. First, her oil industry will provide an increasingly important share of China's energy supplies for her expanding agriculture, industry, and transportation. Second, through the export of crude oil, it will earn the precious foreign exchange so vital for importing foreign technology and equipment. Third, China's petroleum industry itself is in urgent need of modernization so that the country's vast offshore oil deposits can be rapidly exploited.

With the advent of the global energy crisis, China's emerging oil industry is receiving renewed attention. In particular, her policy of undersea oil development is increasingly being highlighted in the Western press. The prevailing optimism that China will seek the participation of major international oil companies in developing her offshore oil seems to be quite warranted. With this in mind, I have compiled this bibliography of sources that can be used to understand the history and development of China's petroleum industry and its probable future trends.

The present work does not cover all information on the petroleum industry in China. Rather, the bibliographies are compiled on the basis of their significance, usefulness for information or research, and availability. Therefore, the user will find works on ancillary topics such as the petrochemical industry and alternate energy re-sources (e.e., shale oil and bio-gas) included in this bibliography, as well as important pieces by specialists such as A.A. Myerhoff and Jan-Olaf Willums, leading geologists on China; Vaclav Smil, an emi-nent geographer; and Professor Chu-yuan Cheng, a prominent Chinese economist.

This annotated bibliography contains a list of approximately 860 entries concerning China's oil industry, from English publications in the main, but augmented by Chinese and Japanese works. It consists of book titles published up to 1980 and serial titles up to the end of 1979. The bibliography is divided into five sections. Section A covers bibliographies, encyclopedias, periodicals, yearbooks, and other English-language reference works. Section B lists books, docu-ments, and monographs published in English. Section C consists of articles in English. Section D includes reference works, books, documents, and monographs published in Chinese and Japanese. Sec-tion E contains Chinese and Japanese articles. Each section is arranged chronologically by date of publication.

For convenience, the East Asian personal names that appear in bibliographical citations have been given in Western order (i.e., first and middle names followed by family names) rather than in East Asian order (i.e., family names followed by first and middle names).

The term "China" used in this bibliography refers to the govern-ment that exercised actual control of the Chinese mainland at the time cited. This will include each Chinese government that con-trolled the mainland before 1949 as well as the present government of the People's Republic of China. Thus, the government of National-ist China (the Republic of China) after 1949 is designated here as the Taiwanese government and listed separately in the subject index under "Taiwan."

The Wade-Giles system is used for the transliteration of Chinese materials. Because the Chinese government has officially adopted Pinyin transcriptions in translation as of January 1, 1979, I have added Pinyin (in parentheses) to Wade-Giles transliterations for Chinese geographical names that appeared in publications after December 1978. A literal translation is provided for Chinese and Japanese titles, corporate bodies, the names of the periodicals, the places of publication, and the publishers. Chinese and Japanese entries are arranged alphabetically by their transliterated titles;

Introduction

Chinese or Japanese titles follow, with English translations in brackets.

There are certain inconsistencies in the English spelling of Chinese geographical names. For example, Pohai Gulf (Bohai Gulf) appears in many writings as Gulf/Bay of Pohai (Bohai) or Chihli. For the sake of uniformity, the term Pohai Gulf is used throughout this bibliography. Annotations retain some traditional spellings (Peking rather than Pei-ching, Kwangtung rather than Kuangtung, etc.).

Like any other bibliography, this work is not perfect. Certain materials may have been omitted because the author is unaware of their existence or availability. Nevertheless, I am confident that this work is the most comprehensive annotated bibliography on China's petroleum industry compiled to date.

I wish to extend my thanks to the librarians and staff at the library of the Mobil Corp. for generously making their facilities available, and also to the librarians at the Engineering Society Library. Special appreciation is due to Mr. Eddie Wang of the East Asian Library at Columbia University, whose bibliographical aid has been instrumental in locating many sources of information. I am also grateful to my colleagues Stephen Harrow, Eric Neubacher, and Spencer H. Means for their interlibrary loan arrangements. I am indebted to Mr. Eiazburo Okuizumi of the East Asia Collection, University of Maryland (College Park), for his advice on Japanese-language entries. Finally, Frank Joseph Shulman, the advisory editor for this series, deserves my sincere appreciation for his meticulous and painstaking work in revising and upgrading this bibliographical guide.

Chronology

1907   Oil flows in commercial quantity at wells in Yench'ang in Shensi (Shaanxi) province.

1936   The discovery of the Yümen (Yumen) oil field in Kansu (Gansu) province heralds the emergence of China's modern petroleum industry.

1943   Production of synthetic fueld from oil shales reaches its peak in Manchuria.

1949   The establishment of the People's Republic of China is announced.

1950   China forms joint ventures with the Soviet Union for exploration of oil and other minerals in Sinkiang (Xinjiang) province.

1955   Karamai oil field begins operation in Sinkiang (Xinjiang) province.

1959   China's leading oil field, Tach'ing (Daqing), is discovered.

1962   China's second largest oil field, Shengli, is opened.

1964   China claims the attainment of self-sufficiency in the production of crude oil.

       Full-scale production of oil commences at the Tach'ing (Daqing) oil field.

1967   Production starts at the Takang (Dagang) oil field.

1968   The construction of China's major petrochemical project, the Peking (later renamed Yenshan [Yanshan]) General Petrochemical Works, begins.

Chronology

1972   China builds her first offshore oil rig, <u>Pohai</u> (Bohai) <u>No. 1</u>, at Talien (Dalian) Shipyard.

1973   China commences crude oil exports to Japan, Hong Kong, North Korea, and North Vietnam.

China completes her key pipeline between the Tach'ing (Daqing) oil field and Ch'inhuangtao (Qinhuangdao), and oil terminal.

1975   Jench'iu (Renqiu) oil field, China's third largest oil field, is discovered in Hopeh (Hebei) Province.

1978   The building of China's north-south oil transmission line, Shantung-Nanking (Shandong-Nanjing, or Lu-ning) pipeline, is completed.

Another major petrochemical complex, the Shanghai General Petrochemical Works, goes on line.

1979   China signs contracts with international oil companies to make seismic surveys in the East China, Yellow, and South China seas.

1980   Elf-Aquitaine, a French oil firm, becomes the first foreign company to drill for petroleum in China's Pohai (Bohai) Gulf.

English Language Publications

(A) Bibliographies, Encyclopedias, Journals, Periodicals, Yearbooks, and Other Reference Works

A1   Oil and Gas Journal. Tulsa, Okla.: Petroleum Publishing Co., 1902-. illus., maps, tables.
     A weekly journal on petroleum and natural gas offering up-to-date information about exploration, drilling, production, processing (refining), and transportation, both at home and abroad, including China. An annual review of worldwide oil and gas appears in its December "Worldwide Report," in which global oil and gas reserves, production, and refining data are covered in capsule form with exclusive field-by-field and plant-by-plant surveys of the world's oil fields and petroleum refineries. In addition, its semiannual construction issue reports on major construction projects in processing and pipeline installations worldwide, including mainland China and Taiwan.

A2   World Oil. Houston: Gulf Publishing Co., 1916-. illus., maps, tables.
     This monthly publication focuses on virtually every aspect of the petroleum industry in China and Taiwan, as well as elsewhere. In addition, its August "International Outlook" examines the development and performance of the oil industry in detailed country-by-country surveys.

A3   Mineral Yearbook. Bureau of Mines. Washington, D.C.: Government Printing Office, 1932/33-. tables.
     This three-volume yearbook provides an annual survey of both domestic and worldwide mineral industries. Volume 3 contains area reports devoted primarily to the mineral industries of China and other countries. A survey of China's mineral fuels, including petroleum and natural gas, appears in the commodity review section, along with trade and production data.

1

A4   <u>Petroleum Economist</u>. London: Petroleum Press Bureau, Ltd.,
      1934-. maps.
         A monthly news service and review of worldwide oil eco-
      nomics and business. It provides up-to-date information about
      oil exploration, drilling, discoveries, reserves, production,
      refining, exports, and the international oil trade. There is
      frequent coverage of oil developments in China. The <u>Petroleum</u>
      <u>Economist</u> is published in English, French, and Japanese edi-
      tions.

A5   <u>Platt's Oilgram News Service</u>. New York: McGraw-Hill, 1934-.
         Platt's provides current, reliable daily news of the
      global oil trade and industry, offering information on oil
      reserves, production, processing, transportation, imports,
      and exports, as well as market reports on the sale and pur-
      chase of oil technology and equipment. There is frequent news
      coverage of China. (Published under the title <u>Oilgram News</u>
      <u>Service</u> until November 1977.)

A6   <u>Far Eastern Economic Review</u>. Hong Kong: Far Eastern Economic
      Review, Ltd., 1946-. illus., maps, tables.
         This weekly journal is an Asian version of <u>Time</u> magazine
      and <u>Business Week</u> combined. The <u>Review</u> focuses on political,
      economic, financial, and business developments throughout
      Asia. There is special emphasis on the Far East, with fre-
      quent feature articles on the petroleum industry in China.
      Since 1974, the annual survey of China's oil industry has been
      incorporated each October in the <u>Review</u>'s yearly "Focus-China"
      section.

A7   <u>Twentieth Century Petroleum Statistics</u>. Dallas: DeGolyer &
      MacNaughten, 1946-. illus., tables.
         This annual publication provides statistical information
      on China's crude oil and natural gas production and reserves,
      refined products demand, oil refinery capacity, and tanker
      fleet size.

A8   <u>Chinese Petroleum Corporation</u>, <u>Annual Report</u>. Taipei:
      1950?-. color illus., maps, graphs.
         A capsule report on all the activities of the oil, gas,
      and petrochemical industry in Taiwan, since the Chinese
      Petroleum Corporation is a government-owned monopoly. It con-
      tains extensive coverage of Taiwan's petroleum and petrochemi-
      cal industries, including onshore and offshore exploration
      (e.g., prospecting and well drilling), production, refining,
      petrochemicals, transportation and storage, marketing, man-
      power development, research and development, and financial
      review. It gives a brief statistical summary of the corpora-
      tion's financial and business operations during the preceding
      ten years and provides the locations of its headquarters and

subsidiaries throughout Taiwan. Major products manufactured
by the company are listed on the final page.

A9   Oil Daily. New York: Oil Daily, Inc., 1951-. illus.
     Published five days a week, this newspaper provides up-to-
date news reports on the world's energy industries. There is
frequent coverage of petroleum and natural gas developments in
China and Taiwan.

A10 China Reconstructs. Peking: China Welfare Institute; dis-
     tributed by Guozi Shudian, 1952-. illus.
     A monthly illustrated journal intended for the general
public. Numerous articles and stories show the progress that
China's economy and industry have made. Includes frequent,
up-to-date reports on the development of the Chinese oil and
gas industry. The journal is published in English, French,
Spanish, Arabic, and Russian editions.

A11 World Petroleum Report. New York: Mona Palmer Publishing
     Co., 1953-1976. illus., maps, tables.
     An annual review of international oil operations and world
petroleum statistical yearbook. In addition to special reports and area studies, its annual national studies cover oil
and gas operations of 170-odd countries on an individual
basis, including China and Taiwan. This publication was ab-
sorbed by International Oil News (entry A25) in 1976.

A12 Peking Review. Peking: Peking Review, 1958-. illus.
     This weekly journal is an official government organ for
informing readers of the latest developments on China's political, economic, and social fronts. There are frequent
feature articles on the progress of China's petroleum and
petrochemical industry. In addition, Peking Review carries
up-to-date information about China's oil industry in the sections entitled "The Week" (changed to "Events and Trends" in
January 1979) and "On the Home Front." (Published under the
title Beijing Review after January 1979.)

A13 China Quarterly. London: Contemporary China Institute of the
     School of Oriental and African Studies, London University,
     1960-.
     A scholarly journal devoted to the study of all aspects of
contemporary China. In its "Quarterly Chronicle and Documentation" section, the journal provides well-documented records
on domestic and international affairs, with frequent reference
to the development of China's petroleum industry and oil
trade.

A14   Petroleum Abstracts. Tulsa, Okla.: Department of Information Science, College of Petroleum Sciences and Engineering, University of Tulsa, 1961-.

    Issued weekly, this journal abstracts U.S. and foreign books, monographs, and articles on petroleum geology exploration, development, production, oil, natural gas, and other energy resources.

A15   Petroleum Geology of Taiwan. Miaoli, Taiwan: Chinese Petroleum Corp., Chinese Petroleum Exploration Division, 1962- (irregular). illus.

    This serial publication is primarily devoted to the geological analysis and hydrocarbon explorations on the island of Taiwan and its vicinity, involving both offshore and onshore tectonic survey and oil and gas drilling activities in the area.

A16   Petroleum Intelligence Weekly. New York: Petroleum & Energy Intelligence Weekly, Inc., 1962-.

    An authoritative weekly on the worldwide oil trade. It provides current information on such developments as the import and export of petroleum and petroleum products, and the sale and purchase of oil technology and equipment among several countries, including China. It is considered a "must" for executives in oil and oil-related businesses around the world.

A17   China Trade Report. Hong Kong: Far Eastern Economic Review, Ltd., 1963-. illus., maps, tables.

    A "comprehensive monthly analysis of market trends and commercial policies of the People's Republic of China," intended to assist businessmen in trading with China. In addition to publishing feature articles assessing and analyzing China's petroleum industry and her foreign oil trade, its "Business Pipeline" section carries up-to-date reports on the fuel situation (primarily oil and gas production and oil trade) and information about China's import of Western oil technology and equipment.

A18   Review of Sino-Soviet Oil. Geneva: Petroconsultants, June 1965-1975?

    This monthly publication "contains unabridged translations and abstracts of articles" published by Russian and East European newspapers and monthly trade journals on oil and natural gas. It covers oil and gas exploration, drilling, reserves, production, transportation, oil trade, and other energy resources in the USSR, China, and Eastern Europe.

A19   Japan Petroleum Weekly. Tokyo: Japan Petroleum Consultants,
       Ltd. (Nihon Sekiyu Konsarutanto), 1966-.
            Provides primary and secondary sources of business infor-
       mation available in English on the development of oil and
       natural gas in China, with often unique reports on Sino-
       Japanese cooperation in the joint exploration of China's off-
       shore oil and gas deposits in the Pohai Gulf.

A20   Economic Reporter (English supp.). Hong Kong: Economic
       Information & Agency, 1967?-. illus.
            This quarterly journal reflects China's official announce-
       ments and reports on her economic and industrial performance
       during each three-month period under review. It frequently
       highlights the progress of China's petroleum and petrochemical
       industry with news about oil and gas exploration, production,
       refining, transportation, and the building of oil fields, re-
       fineries, and petrochemical facilities.

A21   International Petroleum Encyclopedia. Tulsa, Okla.:
       Petroleum Publishing Co., irregular 1967-1972; annual 1973-.
       illus., maps, annual
            Provides a worldwide survey of petroleum reserves, pro-
       duction, and refining capacities on a country-by-country
       basis. Based upon information gathered by the U.S. oil
       industry, it covers both China and Taiwan. Included are a
       world index of oil and gas fields, information about major
       fields, a directory of refineries and petrochemical plants,
       and extensive maps and statistics on worldwide offshore
       operations.

A22   JETRO China Newsletter. Tokyo: Japan External Trade
       Organization, 1968-.
            This quarterly English publication (to be expanded to
       biweekly) provides vital business information on China's
       economy, industry, trade, and government policies, including
       frequent reports on resources such as coal, oil, and gas.
       One of the prime sources on China's energy resources outside
       China in readable English.

A23   Business China. Hong Kong: Business International Asia/
       Pacific, Ltd., 1974-. maps, tables.
            A biweekly report and newsletter on China and other com-
       munist markets in Asia, which supplements the publisher's
       other major newsletter, Business Asia. Its "Pointer" section
       traces industrial trends in China; its "Monitor" section
       closely follows movements in China's industry and frequently
       reports on China's oil industry and trade. This is a useful
       source of information for executives engaged in China trade
       operations.

A24   China Business Review. Washington, D.C.: National Council
       for U.S.-China Trade, 1974-. illus., maps, tables.
            Formerly the U.S.-China Business Review (1974-1977), this
       bimonthly periodical is devoted to U.S. trade with China. Its
       articles frequently cover the current state of China's economy
       and industries (including petroleum and natural gas), as well
       as its oil trade with other countries. In addition, up-to-
       date reports and information on the development of the Chinese
       petroleum industry and Chinese imports of oil technology and
       equipment are included in its "Economic Notes," "Exporter's
       Notes," and "International Notes."

A25   International Oil News. Stamford, Conn.: W.F. Fland Co.,
       1975-.
            Formerly World Petroleum Report (entry A11), this weekly
       news publication tabulates data on international oil and gas
       production, exploration, transportation, and processing,
       including those of China and Taiwan.

A26   Petroleum/Energy Business News Index. New York: American
       Petroleum Institute, Central Abstracting and Indexing Service,
       1975-.
            A monthly publication with annual summaries covering major
       oil journals and dailies, such as Oil and Gas Journal, The Oil
       Daily, The Petroleum Economist, Petroleum Intelligence Weekly,
       and Platt's Oilgram News Service.

A27   Bibliography of Petroleum Geology of Mainland China. Tulsa,
       Okla.: Chenoweth (Philip A.) and Associate, 1976-. 44 pp.
       illus.
            This expensive ($250 per copy) publication is intended for
       business executives who want to determine and identify the
       size and potential of petroleum reserves and possible oppor-
       tunities for their future involvement. It contains over two
       hundred entries, some annotated, on geological aspects of
       China's oil resources. The bibliography is periodically up-
       dated from the original text, which was published in 1973.
       Mention is made of works published by leading international
       scholars about China's oil geology. This publication should
       be of great aid to international oil concerns and oil equip-
       ment manufacturing companies who wish to participate in
       China's oil development.

A28   International Oil Developments: Statistical Survey. Central
       Intelligence Agency. Washington, D.C.: Office of Economic
       Research, April 1976-. illus., tables.
            This biweekly publication contains information on China's
       dialy production of crude oil, natural gas, and other selected
       energy data.

A29   <u>Asian Wall Street Journal</u>. New York: Dow Jones & Co.,
Sept. 1976-. illus.
     This daily version of the <u>Wall Street Journal</u>, is pub-
lished in New York, but issued from Hong Kong. The paper provides extensive coverage on economic, financial, and business
activities in Asia, with particular emphasis on countries in
the Pacific basin, including China. The development of
China's petroleum industry and resources, her energy policies,
and her international energy relations are frequently covered.

A30   <u>China Energy Newsletter</u>. Farmington, Conn.: Eastern Research
Analysis Corp., 1977-.
     This quarterly publication covers a broad spectrum of
China's energy industries ranging from an analysis of Chinese
oil production, refining capacity, oil pipelines, tanker and
terminal construction, and status of new oil fields, to developments in natural gas, coal, hydroelectricity, and nuclear
power. This newsletter also monitors closely changes in the
political and economic developments that affect China's energy
scene.

A31   <u>The Petroleum Industry in China; Its Planning in the Development of Energy Resources: A Bibliography with Selective Anno-
tations</u>, by Raymond Chang. Monticello, Ill.: Council of
Planning Librarians, 1977, 28 pp.
     Provides over 230 bibliographical entries about China's
petroleum and petrochemical industry; lists primary and sec-
ondary sources in Chinese, English, and Japanese, including
reference works, monographs, articles, and documents. Title
arrangement is used for articles, while author entries apply
to books, monographs, and documents. Chinese and Japanese
entries are listed separately and are arranged according to
English romanization. Partially annotated.

A32   <u>China: Facts & Figures Annual</u>, edited by John L. Scherer.
Gulf Breeze, Fla.: Academic International Press, 1978-.
tables.
     This work provides an annual review of developments and
basic data in twelve key sectors in the People's Republic of
China including energy which is further subdivided into production and consumption of primary energy, potential energy
resources, proved oil and gas reserves, production and consumption of oil and natural gas, estimated oil and gas reserves, and oil indicators. In addition, it contains reports
on petroleum-related areas such as merchant tanker fleets,
major petrochemical centers, and oil refineries.

(B)   Books, Documents, and Monographs

B1   Studies on Tectonics and Petroleum in the Yantse (Yangtze)
     Region of Tshung-king (Chungking), by Arnold Albert Heim.
     N.p., n.d.
         One of the earliest geological surveys of the Yangtze
     River valley in the Chungking area of Szechuan province.

B2   Mineral Wealth of China, by William A. Wong. Shanghai:
     Commercial Press, 1927, 129 pp.
         This may be one of the earliest surveys of China's mineral
     resources. It covers the twenty-one major provinces and such
     outer areas as Tibet and Mongolia. The reserves and produc-
     tion of a wide range of minerals such as antimony, asbestos,
     emery, coal, iron ore, mica, lead, gold, silver, mercury, and
     nickel are well recorded and estimated. The provinces that
     the author finds to contain oil deposits and gas reserves
     include Kansu, Shensi, Szechuan, and Yunnan.

B3   Digest of Coal, Iron, and Oil in the Far East, by Boris
     Pavlovich Torgasheff. Honolulu: Institute of Pacific
     Relations, 1929, 63 pp. illus.
         One of the earliest reports on the mineral resources
     (coal, iron, and oil) of East Asia. There is a brief de-
     scription of petroleum in Japan, the Far Eastern part of the
     Soviet Union, China, Taiwan, and the Philippines.

B4   Report on Geological Investigation of Some Oil Fields in
     Sinkiang, by Chi-ching Huang et al. Nanking: Nanking
     Geological Survey of China, 1947, 118 pp. maps.
         Originally published in 1943. This report is the result
     of a survey on stratigraphy, tectonics, the development of
     oil fields, production, refining, and the composition of oil
     in Sinkiang province (in northwest China) conducted by a
     geological survey team sent from Nanking.

B5   General Information of Chinese Petroleum Corporation, by
     National Resources Commission. Taipei: Office of the
     Ministry of Economic Affairs, 1951, 51 pp. illus., tables,
     maps.
         This pamphlet describes the operations and activities of
     the government-owned Chinese Petroleum Corp. in Taiwan. It
     covers the exploration for and production of petroleum and
     natural gas, oil refining, and the oil trade.

B6   Petroleum in Taiwan. N.p., 1955[?], 10 pp. illus.
         A brief introductory pamphlet on the petroleum industry
     and resources of Taiwan in their early stage of post-World
     War II development.

8

B7   <u>Petroleum and Gas Deposits in the Chinese People's Republic</u>,
by Keng Chang. New York: U.S. Joint Publications Research
Service, 1958, 124 pp. illus.

    A translation of reports prepared by the Russians who par-
ticipated in a joint geological survey of petroleum resources
in mainland China during the 1950s.

B8   <u>Communist China's Industry and Materials in Translation:</u>
<u>Petroleum</u>. New York: U.S. Joint Publications Research
Service, 1959, 44 pp.

    A profile of China's petroleum industry and trade through
periodical articles translated from Chinese.

B9   <u>A Compiled Report of 1958 Petroleum Production and Refining</u>
<u>Activities in China</u>. New York: U.S. Joint Publications
Research Service; distributed by the Office of Technical
Services, Department of Commerce, Washington, D.C., 1959,
52 pp.

    A collection of articles, translated from various Chinese
periodicals, dealing with the operation and performance of
China's oil industry. Crude oil production and refining
during 1958 are emphasized.

B10   <u>Research, Production, and Utilization of Petroleum in Com-</u>
<u>munist China</u>. New York: U.S. Joint Publications Research
Service, 1959, 50 pp.

    Primarily an examination of China's oil situation, (such
as demand and supply) and the technological level of China's
petroleum industry. All information is translated from the
January-March 1958 issues (nos. 1-3) of China's leading oil
journal, <u>Shih-yu lien chi</u>.

B11   <u>Communist China's Petroleum Situation</u>, by Kung-chia Yeh.
Santa Monica, Calif.: Rand Corporation, 1962, 69 pp. maps,
tables. (Rand Memorandum RM-3160-PR.)

    Critically evaluates China's petroleum production, con-
sumption, and imports during the 1949-1960 period. The author
estimates China's total oil supply in 1960 to be around 5 million tons, with petroleum imports coming primarily from the
Soviet Union. He forecasts that prospects for greatly expanded domestic production are not bright and that China is
likely to continue importing oil from the Soviet bloc for some
time to come.

B12   <u>Economic Development and Use of Energy Resources in Communist</u>
<u>China</u>, by Yuan-li Wu, with the assistance of H.C. Ling. New
York: Praeger, for the Hoover Institution on War, Revolution,
and Peace, 1963, 275 pp. illus., maps, tables, graphs.

    The first well-documented, scholarly work on China's
energy resources covering the years 1949-1960. Emphasis is
placed on the development of China's energy resources in

relation to the country's other economic activities. Sizable
data on China's coal and electricity have been amassed. The
petroleum industry is covered in only a small portion of the
work, since it did not play a significant role in the energy
picture between 1949 and 1960. Four aspects of the infant
petroleum industry are studied: the size and distribution of
oil reserves, the producing fields and refineries, the production of crude oil, and petroleum imports. Bibliographical
notes and statistical tables are extensive.

B13   The Survey of the Energy Economy of Taiwan. Washington, D.C.:
China Energy Resources Study Team, 1965, 219 pp. illus., map.
      An extensive examination and analysis of Taiwan's power
resources including coal, electric power, oil, and natural
gas.

B14   An Economic Geography of China, by Thomas R. Tregear. London:
Butterworth; New York: American Elsevier, 1970, 275 pp.
illus., maps.
      Describes and analyzes the country and economy of China
from a geographical standpoint, with special reference to the
development and changes that have occurred since 1949. Traces
the economic growth of China, her geographical makeup, contemporary agriculture and industrial and transportation development. A brief account is provided of the historical development of oil and gas in China and post-1949 Soviet assistance.
Estimates that 90% of China's oil reserves lie in the northwest, in such oil fields as Karamai, Yümen, Tsaidam, Szechuan,
and Tach'ing.

B15   "Investment in Taiwan: A Look at the Petrochemical Industry
and United States Involvement in its Growth and Development,"
by Kong-heong Tan. M.S. thesis, Massachusetts Institute of
Technology, 1971, 130 pp.

B16   China's Changing Map: National and Regional Development,
1949-71, by Theodore Shabad. Rev. ed. New York: Praeger,
1972, 370 pp. maps, tables.
      Examines the geographical changes which have occurred on
China's political and economic map since 1949. Emphasis is
placed on recent developments in China's natural resources
base, including her energy resources. China's petroleum industry is discussed in the section on China's national economy
as well as in the regional breakdown. Thus, the Tach'ing oil
field is described in the section on Manchuria, the Shengli
oil field in Shantung province, the Karamai oil field in
Sinkiang province, the Yümen oil field in Kansu province, the
Lenghu oil field in Tsinghai province, natural gas fields in
Szechuan province, and finally, oil shale in Kwangtung and
Manchuria. This book provides very informative sketches of
the geographical distribution of China's resource industries.

B17   <u>Taching: Red Banner on China's Industrial Front</u>. Peking:
Foreign Language Press, 1972, 46 pp. illus.
    A pamphlet issued by the Chinese government describing the
development of the Tach'ing oil field. It stresses the spirit
of self-reliance through which the Chinese built a vast oil
base entirely on their own.

B18   <u>The Outlook for the Petrochemical Industry in Taiwan, June
1973</u>. Taipei: Industrial Development and Investment Center,
Ministry of Economic Affairs, Republic of China, 1973, 107 pp.
illus.
    One of the most complete analyses and appraisals of the
past, present, and future development of the petrochemical and
ancillary industries in Taiwan. Envisions a continued trend
of rapid growth and diversification of the industry, accom-
panied by heavy capital investment in the industry.

B19   <u>PRC Oil: For the Lamps of China?</u> Washington, D.C.:
Department of State, Bureau of Intelligence and Research,
1973, 5 pp.
    A brief survey and review of China's oil situation from
1970 to 1972, covering exploration, production, refining,
transportation, and consumption. The report views China as
being basically self-sufficient in oil at her current low
level of consumption.

B20   <u>The Taching Oilfield: A Maoist Model for Economic Develop-
ment</u>, by Leslie W. Chan. Canberra: Australian National
University Press, 1974, 28 pp. illus., maps. (Contemporary
China Papers, no. 8.)
    Analyzes the Tach'ing oil field from socioeconomic angles
rather than dissecting it as an industrial entity. Follows
the developmental history of this oil base from its discovery
through the hiatus during the Cultural Revolution up to 1973.
Evaluates Tach'ing's success and failures, and examines the
feasibility of Tach'ing as a a national model for China's
economic development. The author seems to conclude that
Tach'ing is a very unique case of industrial enterprise combined with agricultural settlements and subsidized heavily by
the state. Thus, it might not be possible for the experiment
to be repeated elsewhere in China.

B21   <u>China</u>, by Choon-ho Park. Newark, Del.: Center for the Study
of Marine Policy, College of Marine Studies, University of
Delaware, 1975, 63 pp. illus., maps.
    Presents a brief historical review and analysis of the
development of China's energy resources and her energy policies, with special emphasis on petroleum. Park examines the
potential of China's current resources and relates state
policy and government structure to the allocation of oil resources for domestic use and export. He also estimates

China's oil reserves, production, and refining capacity (including a breakdown by individual oil fields) and discusses the demand and supply factors that will affect Chinese oil exports in the 1970s and 1980s. Finally, Park concludes that while China's basic policy for the development of oil resources is self-reliance, there is room for flexibility where future foreign participation in China's energy ventures is concerned.

B22 China: Energy Balance Projections. Washington, D.C.: Central Intelligence Agency, 1975, 33 pp. illus.

Attempts to estimate China's current energy output and consumption and to forecast their future growth. Also tries to examine China's major energy sources, to identify energy-consuming sectors within the Chinese economy, to predict energy balances in 1980 and 1985, and to estimate crude oil export volumes in 1980 and 1985. Through an analysis of demand and supply factors, the CIA concludes that the rapidly expanding domestic need for energy will most likely limit China's crude oil exports to under 1.3 million b/d in 1985 and will probably keep China from becoming a major factor in the world oil market during the 1980s.

B23 China Trade Guide, by J.E. Metcalf and V.K. Rangathan. New York: First National City Bank, 1975, 76 pp. illus. (some color), maps.

Primarily studies China's import market and export prospects. Her petroleum industry, however, is introduced and assessed as offering a good prospect for both the import of oil technology and the export of crude oil. In chapter 5, the authors briefly survey China's oil industry, examining in particular her oil refineries, oil technology, and petrochemicals.

B24 "China's Offshore Oil: Application of a Framework for Evaluating Oil and Gas Potentials under Uncertainty," by Jan-Olaf Willums. Ph.D. dissertation, Massachusetts Institute of Technology, Department of Ocean Engineering, 1975, 460 leaves. illus., maps.

Probably one of the most exhaustive and systematic analyses of hydrocarbon deposits in China's offshore waters. Willums estimates that China should be able to produce 43 and 154 million tons of crude oil from her offshore oil fields in 1980 and 1985, respectively, if she can maintain political stability, adequate financial strength, and technological capability in offshore logistics. He forecasts a growing export of China's crude oil to Japan by 1985 and also a promising market for offshore oil technology and equipment. Very optimistic about China's oil production from her offshore resources. (Copies available through the Microreproduction Laboratory, M.I.T.)

B25   <u>China's Oil Industry: A Background Survey</u>. London: Sino-
British Trade Council, 1975, 48 leaves. illus., maps.
    Originally intended as workbook at the Conference on
China's Oil and Trading Possibilities, held in Glasgow on
June 26, 1975 and organized by the Sino-British Trade Council.
This book covers China's oil resources, production, refining,
transportation, exports, and the evaluation of China's market
for oil-related equipment. Predicts China's growing oil output from both onshore and offshore oil fields and a good gain
in crude oil exports by 1980.

B26   <u>Mineral Resources of China</u>, by A.B. Ikonnikov. Boulder,
Colo.: Geological Society of America, 1975, 6 microfiche
sheets (562 pp.). maps.
    Presents a comprehensive geological survey of China's
mineral resources developed since 1949. Ikonnikov provides
a sketch of China's structural geology including geotectonics
and tectonic units. He depicts major mineral deposits--coal,
oil and gas, iron ore, heavy metals, light, and rare metals.
In addition, Ikonnikov presents an informative analysis of
oil and gas, plus oil shale, covering the general background,
surveying the oil-bearing and oil shale regions, and apprais-
ing the hydrocarbon reserves of China. Also included is an
index of Chinese geographic names.

B27   <u>The People's Republic of China: A New Industrial Power with
a Strong Mineral Base</u>, by Kung-ping Wang. Washington, D.C.:
Government Printing Office, 1975, 96 pp. illus., maps.
    Reviews the development of the mineral industry in China.
Drawing information from commercial circles, technical journals, international trade publications, Japanese sources,
visitors, general literature, and the Chinese press, Wang
points out the world significance of China's minerals, traces
the history of their exploitation, assesses supplies, and discusses regional factors affecting mineral development. A
sectoral evaluation is made of major mineral industries such
as coal, power, iron and steel, oil and gas, nonferrous
metals, fertilizers, and chemicals. With regard to the oil
and gas industry, he specifically reviews its output, production potential, offshore exploration, refining, petrochemicals, exports, consumption, and the major oil fields. Mention
is also made of shale oil resources in Kwangtung province and
Manchuria. (For Wang's subsequent assessments see entries B42
and C414.)

B28   <u>The Petroleum Industry of the People's Republic of China</u>, by
H.C. Ling. Stanford, Calif.: Hoover Institution Press, 1975,
264 pp. illus., maps, graphs.
    Presents a broad examination of China's crude oil supply
and demand and evaluates major factors affecting it: China's
institutional framework, petroleum resources, refining and

technology, and transportation of oil. Despite these constraints, the author concludes China has adequate oil deposits
to meet her demand and has surpluses for export. But if demand for oil grows more rapidly than in the past, and if problems of technology and transportation cannot be surmounted,
the rate of growth in oil production would be less rapid than
expected.

B29   China's Energy: Achievements, Problems, Prospects, by Vaclav
      Smil. New York: Praeger, 1976, 275 pp. illus., maps, tables.
         The first comprehensive survey of China's energy resources
      (1949-1975) since the 1963 publication of Wu Yuan-li's
      Economic Development and the Use of Energy Resources in Com-
      munist China (entry B12). It assays China's energy potentials
      on the basis of past growth, current trends, and future course.
      It also analyzes China's fossil fuels (coal, oil and gas),
      hydroenergy, and energy trade. Drawing on primary sources, as
      well as Japanese, Soviet, and Nationalist Chinese studies, the
      author amplifies his earlier monograph "Energy in China"
      (entry C311), which appeared in the March 1976 issue of the
      China Quarterly. Concludes that China's energy potential--
      particularly in the areas of oil production and trade--is very
      promising.

B30   China's Energy Policies and Resource Development, edited by
      Thomas Finger. Stanford, Calif.: U.S.-China Relations Pro-
      gram, Stanford University, 1976[?], 56 pp. (U.S.-China Rela-
      tions Report, no. 1.)
         A collection of papers examining China's domestic and
      international energy policies and the exploitation of her
      energy resources. Examines the technical and economic con-
      straints that affect present and potential production of
      petroleum and natural gas.

B31   China's Petroleum Industry. Washington, D.C.: National
      Council for U.S.-China Trade, 1976, 122 pp. maps.
         Originally intended as a workbook at the Conference on
      China's Petroleum Industry, held in Houston on June 23, 1976
      and cosponsored by the National Council for U.S.-China Trade
      and the Houston Chamber of Commerce. This publication is
      probably one of the most informative and valuable sources on
      the current state of the oil industry in China, her oil export
      potential, and her imports of oil-related equipment. It is
      also aimed at international business executives who want to
      size up China's market for petroleum-related facilities and
      oil technology, assess the international competition, and
      determine whether significant opportunities exist for the
      sale of petroleum equipment to China. Accordingly, this pub-
      lication provides information on China's petroleum reserve
      estimates, oil production, refinery capacities, oil infra-
      structure, oil exports, and her oil equipment industry. It

also furnishes data based on exhibits in China of foreign
petroleum equipment and sales of oil-related equipment to
China by American and international companies. China's
Petroleum Industry should also interest scholars, researchers,
and economic analysts who want basic facts about China's
petroleum industry.

B32   China's Petroleum Industry: Output Growth and Export Potentials, by Chu-yuan Cheng. New York: Praeger, 1976, 244 pp.
illus., maps, tables. (Praeger Special Studies in International Economics and Development.)
     Represents the culmination of years of study by the
author. A realistic evaluation of the development of China's
petroleum industry, including estimates of China's crude oil
output and reserves and a comprehensive survey of major
fields, refineries, and transportation facilities. Special
attention is given to China's oil equipment industry, her
import of oil technology, and the contribution of the oil
industry to China's national economy. Finally, the author
assesses China's potential oil exports objectively. An extensive bibliography is included.

B33   Chinese Oil: Development Prospects and Potential Impact, by
Randall W. Hardy. Washington, D.C.: Center for Strategic
and International Studies, Georgetown University, 1976,
[88] pp. maps, tables.
     The author, a former White House aide who might have had
access to some classified materials, assesses China's oil
prospects. He evaluates China's oil reserves, problems of
marketing and distribution, and offshore developments. He
also appraises China's domestic "variables" and its foreign
trade. Peking's oil diplomacy in its relations with Japan
and the USSR and concomitant questions are discussed. In con-
clusion, Hardy predicts that China will produce 4 million b/d
of crude oil by 1980 and 8 million b/d by 1990 if she has the
political stability, financial strength, and the technological
capability to surmount present constraints. The author, however, foresees certain limits on exports because of rising
domestic consumption. (See entry B46 for a later edition of
this work.)

B34   Energy Policies of the World, edited by Gerard J. Mangone.
New York: Elsevier, 1976-. illus., maps.
     Volume 1 of this series reviews and examines energy re-
sources such as petroleum, natural gas, coal, hydropower, and
other alternate energy resources in Canada, China, Iran,
Venezuela, and the Arab states. Choon-ho Park is the con-
tributing author on China. Park studies the potential of
China's oil resources and relates her policy in allocating
hydrocarbon for both internal use and export purposes. In
addition, he discusses China's oil reserves, exploration,

production, and refining capacity, as well as demand and supply factors that will ultimately affect Chinese oil exports in
the late 1970s and the 1980s. In volume 3 (1979), contribut-
ing author John Franklin Copper provides extensive information
on Taiwan's energy situation, including her historical use of
energy, sources of energy, import of petroleum, and alternate
energy resources. Particular emphasis is placed on Taiwan's
domestic production of natural gas and the problem of huge
petroleum imports, which are Taiwan's most costly import item.
For Taiwan's energy future, Copper weighs the possibility of
new sources of alternate energy, such as geothermal power and
solar energy in southern Taiwan.

B35   Implications of Prospective Chinese Petroleum Development to
      1980, by Peter W. Colm, Rosemary Hayes, and Edwin Jones.
      Arlington, Va.: Institute for Defense Analysis, International
      and Social Studies Division, 1976, 73 pp. illus., maps,
      tables.
          Offers a broad survey of the development of China's
      petroleum industry and critically dissects various estimates
      of China's oil reserves, production, refining capacity,
      domestic consumption, and export potential through 1980. The
      projected performance of China's oil industry takes into
      account divergent views of the country's oil prospects given
      by a group of oil analysts with different backgrounds. This
      report also focuses attention on the oil industry's contribu-
      tion to China's domestic economy and its role in cementing her
      new ties with Japan and other Asian countries.

B36   "The International Energy Policies of the People's Republic of
      China," by Kim Woodard. Ph.D. dissertation, Stanford Univer-
      sity, 1976, 476 pp. illus., tables.
          An exhaustive empirical investigation of China's inter-
      national energy policies toward Asian nations, the devel-
      oped countries, and the Third World. It first assesses
      China's domestic energy resources, production, consumption,
      and organizational structure. It then relates them to the
      process of decision-making with respect to her international
      energy policies and to the role China might play in the world
      oil market. In this penetrating work Woodard observes that a
      variety of factors tend to constrain China from playing a sig-
      nificant role on the international energy scene. Two of these
      are (1) China's large domestic energy consumption potential,
      which will place a ceiling on her surplus of oil, and (2) the
      possible depletion of China's onshore oil deposits if she
      fails to tap her vast offshore oil reserves on a large scale
      by 1985.

B37   <u>Oil and Asian Rivals: Sino-Soviet Conflict; Japan and the Oil</u>
<u>Crisis</u>. U.S. Congress, House Subcommittee on Asian and
Pacific Affairs. Washington, D.C.: Government Printing
Office, 1976, 476 pp. illus.
      Hearings examining Sino-Soviet relations and the impact of
the energy crisis on Asian countries. Among the presentations
was Choon-ho Park's article "Oil Under Troubled Waters: The
Northeast Asia Sea-Bed Controversy," which described the
international dispute over China's supposedly oil-rich outer
continental shelf. In addition, Park read a statement on
China's petroleum industry, covering her oil resources, production, and supply and demand situation (pp. 41-51). A discussion about China's oil followed this statement.

B38   <u>Winds of Change: Evolving Relations and Interests in South-</u>
<u>east Asia</u>, by Mike Mansfield. Washington, D.C.: U.S. Govern-
ment Printing Office, 1976, 47 pp. map (fold).
      Senator Mike Mansfield's report on his official tour of
Southeast Asian nations, with major emphasis on the development of petroleum resources and potentials of the individual
countries, including China.

B39   <u>China, Oil and Asia: Conflict Ahead?</u>, by Selig S. Harrison.
New York: Columbia University Press, 1977, 317 pp. plates,
illus., maps.
      An amplification of the author's challenging article
"China: The Next Oil Giant; Time Bomb in East Asia," published in 1975 (entry C280). Reports on China's oil potential, particularly her offshore reserves, and discusses
possible controversies arising out of boundary disputes with
Japan, Taiwan, South Korea, Vietnam, and more remotely, the
USSR. Harrison urges the U.S. to avoid becoming entangled in
disputes over the continental shelf. He predicts that China
will produce 400 million tons of oil a year by 1990 if offshore deposits are developed, and foresees a rise in the Sino-
Japanese oil trade. Finally, Harrison cautions that China's
surplus oil for export might not be large because of her huge
domestic consumption.

B40   <u>China Oil Production Prospects</u>. Washington, D.C.: Central
Intelligence Agency, 1977, 28 pp. illus., maps.
      Critically examines China's total oil reserves, including
her onshore, offshore, and oil shale reserves, and covers
crude oil production, major fields, and potential output.
China's onshore oil reserves are estimated at 39 billion
barrels, and the combined total of offshore and onshore reserves at about 80 billion barrels. China's oil output in
1976 is calculated to be 1.7 million b/d and to be increasing
at an annual rate of 20%. The CIA forecasts a rise in output
to 2.4-2.8 million b/d in 1980 and a surplus of oil for
export that year within the range of 200-600,000 b/d. It

concludes that expanding domestic consumption will absorb
total producing capacity within a decade unless Western oil
fields and offshore deposits are developed more rapidly.

B41   The Chinese Petroleum Industry: Myth and Reality, by Bruce J.
      Esposito. Washington, D.C.: Department of State, 1977, 7 pp.
      (FAR paper no. 28177.)
           Originally presented at the New England Conference of the
      Association for Asian Studies, Amherst, Mass., October 16,
      1977. This discussion paper provides a well-balanced review
      of the extent of China's oil reserves, the Chinese capacity
      for crude oil production, prospects for oil exports (with
      particular regard to Japanese market), and problems for future
      oil development.

B42   Mineral Resources and Basic Industries in the People's Repub-
      lic of China, by K.P. Wang. Boulder, Colo.: Westview Press,
      1977.
           An updating and adaptation of the author's 1975 work, The
      People's Republic of China: A New Industrial Power with a
      Strong Mineral Base (entry B27). This edition offers more
      extensive coverage of oil and gas within the context of
      China's overall mineral resources. Offshore exploration and
      the construction of pipelines and tanker ports are dealt with
      in some detail, and information about China's petrochemical
      industry is brought up to date. Wang compares various esti-
      mates made of China's petroleum output, potential, refining
      capacity, consumption figures, and export volume with his own.
      Finally, he terms China's oil position medium-sized and pre-
      dicts that China has a reasonably good chance of joining the
      league of the world's big five oil producers.

B43   Oil in the People's Republic of China: Industry Structure,
      Production, Export, by Wolfgang Bartke. Montreal: McGill-
      Queens University Press, 1977, 125 pp. illus., maps.
           An English translation of Die Ölwirtschaft der Volks-
      republik China, published in 1975 (the British edition is
      published by C. Hurt & Co. of London). Provides a summary of
      China's oil industry and breaks down production by geographi-
      cal regions, refineries, and vitas of China's leading cadres
      in the industry. The inclusion of geographical and personal
      names in Chinese should also be helpful to students of China's
      petroleum industry.

B44   South China Sea Oil: Two Problems of Ownership and Develop-
      ment, by Roderick O'Brien. Singapore: Institute of Southeast
      Asian Studies, 1977, 85 pp. map.
           Originally written as an M.A. thesis at the University of
      Hong Kong. Tries to analyze the geological and physiographi-
      cal significance of oil deposits in the South China Sea, which
      cover a broad offshore area stretching from southern China and

Taiwan to northern Borneo. Also dissects the problems arising
from the claims on the ownership of petroleum resources in the
region, examining the issue of economic nationalism and the
patterns of oil development in the ASEAN countries.

B45   China and United States Policy: Report of Senator Henry M.
Jackson to the Committee on Armed Services and the Committee
on Energy and Natural Resources, U.S. Senate, March 1978.
Washington, D.C.: Government Printing Office, 1978, 8 pp.
     Deals with China's energy policy, with special emphasis on
the state of its oil industry. The report urges the U.S. government to help China develop its vast offshore oil deposits.

B46   China's Oil Future: A Case of Modest Expectations, by Randall
W. Hardy. Boulder, Colo.: Westview Press, 1978, 148 pp.
graphs, maps.
     An updating, modification, and amplification of the au-
thor's 1976 work, Chinese Oil: Development Prospects and
Potential Impact (entry B33). In this new edition, Hardy
offers a more modest estimate of China's future oil production
and views China's oil export potential as limited. It will be
very difficult for China to produce enough oil to satisfy her
expanding domestic demand and to have large surpluses for export. China also has to overcome political, economic, and
technological problems to achieve her goal. This book is an
attempt to examine these problems, to evaluate the possibility
of resolving them, and to assess the role China is likely to
play as a major oil producing country in the 1980s.

B47   Doing Business with the People's Republic of China: Industries and Markets, by Bohdan O. Szuprowicz and Maria R.
Szuprowicz. New York: John Wiley, 1978, 449 pp. illus.,
maps.
     An up-to-date business manual for American exporters and
importers who want to trade with China. Provides information
about the size of China's industries, current import markets,
and future potential. In the section on energy and fuels, the
authors provide the latest available information about China's
energy reserves and resources and furnish current data on
China's petroleum production and consumption, including a
breakdown on individual fields, the current status of pipelines, and the capacity of refineries. Emphasis is also
placed on the significance of China's crude oil exports. The
extent of the authors' market research is well illustrated by
their inclusion of lists of major Chinese imports of petroleum
equipment coupled with contracts for basic petrochemical turn-
key plants from Western countries between 1963 and 1977. The
book also provides a good profile of China's rapidly expanding
petrochemical and natural gas industries.

B48   Impressions of Taching Oil Field, by Shan-hao Chiang. Peking:
       Foreign Languages Press, 1978, 46 pp.
           A collection of articles on the Tach'ing oil field that
       originally appeared in installments in Peking Review before
       1977.

B49   The Offshore Oil and Gas Field Equipment Market in the Soviet
       Union and People's Republic of China. New York: Frost &
       Sullivan, 1978, [321] pp.
           An authoritative market survey of oil and gas equipment
       markets in China and the USSR, with particular emphasis on the
       technology and equipment China needs in developing her off-
       shore oil and gas.

B50   China's Oil: Problems and Prospects, by Sevinc Carlson.
       Washington, D.C.: Center for Strategic and International
       Studies, Georgetown University; distributed by McGraw-Hill,
       1979, 128 pp. illus., maps, tables.
           Presents a critical analysis of the future of the petro-
       leum and natural gas industry in China. Appraises China's
       plans for future oil production and development, her oil and
       gas reserves, future export plans and potential, and con-
       straints in transportation and oil technology. Also included
       are aspects of China's alternate energy resources such as
       shale oil, coal, hydroelectricity, and nuclear energy.

B51   Oil & Chemicals in China. Cleveland, Ohio: Predicasts, 1979,
       81 leaves. (Regional Market Report RG-10.)
           Reports the result of marketing research and market sur-
       veys of the chemical and petroleum industry in China, con-
       ducted by Predicasts.

B52   China's Petroleum Organization and Manpower, Including Dis-
       semination of Technology, by Jeffrey Schultz. Washington,
       D.C.: National Council for U.S.-China Trade, 1980, 108 pp.
       illus., tables, maps.
           Highlights the horizontal and vertical structures of
       China's petroleum industry, including oil production units and
       their roles in technology dissemination. Evaluates China's
       oil and gas fields, reviews the development of her offshore
       oil, and provides detailed description of the industry's
       managerial and technical manpower at different levels of orga-
       nization. Includes a wall chart outlining the industry struc-
       ture and identifies 400 key personnel and their positions.
       Expensive as this book is ($200 a copy), it is a vital tool
       for anyone intending to do business with the Chinese petroleum
       industry.

B53   The International Energy Relations of China, by Kim Woodard.
       Stanford, Calif.: Stanford University Press, 1980, 717 pp.
       charts, maps, tables.
            Updates and amplifies the author's thesis, "The Inter-
       national Energy Policies of the People's Republic of China"
       (entry B36). China's domestic energy system and her energy
       policies in relations with other countries are carefully ex-
       amined and analyzed, with particular emphasis on petroleum
       resources. Part 1 of the work plots the course of China's
       international energy policy from 1949 to the end of this
       century. Part 2 covers China's energy balance sheet in the
       form of nearly 100 statistical tables and computerized pro-
       jection models that project individual segments of China's
       energy industry to the year 2000. This book is a major con-
       tribution to the study of China's energy industry and her
       international energy trade.

(C) Articles

C1 "Explorations in China," by Myron L. Fuller. <u>Bulletin of the</u>
<u>American Association of Petroleum Geologists</u> 3 (1919):99-116.
      The first report on Fuller and Clapp's geological expedi-
tions in North China during 1913-1915, in which they directed
six geological parties traversing 20,000 miles through the
provinces of Chihli, Shansi, Shensi, Kansu, and northern
Honan searching for petroleum resources.

C2 "Oil Prospects in Northeastern China," by Myron L. Fuller and
Frederick G. Clapp. <u>Bulletin of the American Association of</u>
<u>Petroleum Geologists</u> 10, no. 11 (1926):1073-1117.
      Probably the earliest and most comprehensive published
report on China's petroleum prospects by men of extensive
experience in petroleum geology. The authors' expeditions
covered the provinces of Chihli, Shansi, Shensi, Kansu, and
northern Honan, some 20,000 miles. Initial information indicated that China was not rich in petroleum resources. Except
for the provinces of Kansu and Shensi, there appeared to be no
prospect of large-scale commercial production of crude oil in
northern China.

C3 "Petroleum--China." In <u>Ores and Industry in the Far East</u>, by
H. Foster Bain, New York: Council on Foreign Relations, 1933,
pp. 123-33.
      Probably one of the earliest geological reports on China's
petroleum resources made by Western geologists. However,
their initial survey did not indicate any major oil findings
of commercial significance in China proper, Manchuria, and
Inner Mongolia, except for the provinces of Kansu and Shensi,
and some oil shale prospects at Fushun, Manchuria.

C4 "China's Future Producing Possibilities More Promising Than
Developments Indicate," by C.Y. Hsieh. <u>Oil Weekly</u> 80
(Dec. 30, 1935):20-30. illus., maps.
      One of the earliest, best documented analyses and reports
of China's oil prospects. Professor Hsieh of Peking University (director of the Geological Survey of China) and his team
made an extensive prospecting mission to remote basins in
Sinkiang, Szechuan, northern Shensi, and Kansu, where they
surveyed and analyzed the stratigraphic and structural co-
relations with unsophisticated equipment in the 1920s and
early 1930s. Even though he did not make any major oil find
at the time, Hsieh seems to have been optimistic about the
future discovery of oil deposits and the development of oil
shale in many parts of China.

C5   "Search for Oil in Manchu Kuo." <u>Far Eastern Engineer</u> 12
(Dec. 1940):805-25. illus.
      Describes the Japanese geological survey in Manchuria for
oil resources and the Japanese development of the region's
shale oil industry.

C6   "China's Production of 3,000 Barrels Daily Starts the Nation
toward Self-Sufficiency," by John P. O'Donnell. <u>Oil and Gas
Journal</u> 43 (March 9, 1944):34-35. illus.
      An oil field in Kansu, 900 miles north of China's wartime
capital Chungking, has attained a daily production of 3,000
barrels. The development of the oil field has been made possible by the aid of American oil men and equipment. The difficult terrain makes it very hard to transport equipment to
the site of operation. Compared with the U.S., oil output is
very small, but this might be the maximum output possible at
this site.

C7   "China Is Awake to Its Petroleum Possibilities." <u>Oil Weekly</u>
116 (Dec. 11, 1944):2, 130.
      One of the earliest reports available on the activities of
the petroleum industry in China. Reports that because of the
Japanese occupation of China's vast east coast, oil production
in free China is confined to Kansu and Sinkiang province in
the northwest. In Kansu, 14 oil wells produce 2,500 b/d of
crude oil, only half of their potential. A small refinery
nearby turns out gasoline for local use. Meanwhile, the
Soviets are helping the Chinese explore and drill in the area,
and there are reports of the opening of a new oil field.

C8   "Petroleum in Kansu Province, China," by Martin J. Gavin.
<u>Petroleum Engineer</u> 17 (Oct. 1945):181-82, 184. illus.
      Reports on the discovery and development of the first com-
mercial oil field in China, at Yümen. The coastal blockade by
Japanese invading forces makes it imperative to find and produce fuel oil in China's interior. Under such circumstances,
this new oil base has been built on difficult terrain. A
small refinery without cracking facilities produces straight-
run gasoline, kerosene, gas oil, and wax, which have to be
transported by trucks and rafts to the city of Chungking for
consumption.

C9   "Oil and China's Future," by L.J. Logan. <u>Oil Weekly</u> 120
(Dec. 3, 1945):37, 39-43.
      A general survey and evaluation of China's petroleum re-
serves, production, and prospects at the end of World War II.
The author regards China as a country with good oil potential
and discusses the possibility of U.S. financial and technological assistance in developing China's oil reserves.

C10   "War's End Brings Accelerated Activity in Yumen, China's Only
       Oil Field," by K.C. Lu. Oil and Gas Journal 44 (Dec. 29,
       1945):34, 253-257. illus., map.
           Describes the Yümen oil field in its early stage of devel-
       opment under the Nationalist government at the end of World
       War II. Discovered in 1939, this oil base has only seven
       wells drilled with a total output of 1,500-2,200 b/d in 1942.
       A small refinery is being built. This oil field is located in
       Kansu province about 450 miles from the city of Lanchou and
       2,000 miles from the east coast. It already employs about
       7,000.

C11   "China's Oil Policy," by K.C. Lu. Oil Weekly 121 (May 6,
       1946):27. illus.
           Written by a Chinese petroleum engineer while being
       trained in the U.S. Explains China's postwar economic policy
       regarding foreign investment and participation in the manage-
       ment of Chinese companies. Lu states that postwar China needs
       financial and technological assistance from abroad, particu-
       larly from the U.S., and that there is no tax discrimination
       against foreign capital investments in China.

C12   "Yenchang Oil Field," by Joe R. White. Petroleum Engineer
       21 (June 1949):6, 38, 40, 42-46, 48. illus.
           A portrayal of the oldest oil field in China, located in
       the northeast. The first strike at Yenchang was made by a
       Japanese engineer in 1908. The oil field consists of eight
       wells, (five drilled by Socony-Vacuum), producing about
       100 b/d. A small refinery has the capacity of 320 b/d, and
       one of its main functions is to purify paraffin to make
       candles. The field is noncommercial by U.S. standards.

C13   "China: 1950 Production of Oil." World Oil 133 (July 15,
       1951):242.
           China's crude oil production in 1950 is estimated at
       803,000 barrels--a gain of 73,000 barrels over 1949. A joint
       Sino-Soviet company has been formed to engage in oil explora-
       tion in Sinkiang province. Meanwhile, U.S. oil interests in
       China, such as Esso and Caltex have been placed under state
       control.

C14   "Oil Industry of China," by John Cardew. Petroleum Times 56
       (Mar. 21, 1952):222-26. illus.
           A profile of China's oil industry in its infancy. De-
       scribes China's effort to develop her oil deposits in north-
       western provinces because of the allied embargo on sales of
       petroleum products to China during the Korean War. Reports on
       China's success in restoring its oil production to the pre-
       World War II level in 1950. Also reports on Soviet, Polish,
       and Czechoslovakian assistance in building the oil industry in
       China.

C15   "China Is Rich in Oil." China Reconstructs 1, no. 2
      (Mar./Apr. 1952):18-19. illus.
         Tries to refute the assertion that China is an oil-poor
      nation. Maintains that China has great petroleum resources
      and that facts and figures point to rapidly developing oil-
      extracting and refining industries within the country.

C16   "China: 1951 Activities." World Oil 135 (July 15, 1952):216.
         China's crude oil production in 1951 slipped to 728,000
      barrels from 803,000 in 1950, a 3% drop. The U.S. and Great
      Britain staged an oil embargo against China because of the
      Korean War. Known oil reserves in 1951 are estimated to be
      1.5 billion barrels. Most oil shale plants built by the
      Japanese have been restored to operations.

C17   "Karamai--China's New Oil Town," by Hsu Chien. Eastern World
      11 (Oct. 1957):37-38.
         Describes the development of an oil field and the building
      of a new town in the province of Sinkiang in the 1950s.

C18   "China's Sky-Rocketing Oil Output," by Chi-lin Chu. Peking
      Review 1, no. 23 (Aug. 5, 1958):9-11. graphs.
         China's crude oil production increased more than tenfold
      from 1949 to 1957, from a mere 122,000 tons a year to
      1,460,000 tons.

C19   "Taiwan's Quest of Oil," by Godwin Chu. Free China Review 8
      (Oct. 1958):22-25.
         Chu describes extensive wildcat drilling by the China
      Petroleum Corporation on Taiwan's west coast in search of oil
      and gas. The results have been mixed.

C20   "Behind the Bamboo Curtain: An All-out Drive to Boost Out-
      put." Petroleum World 7 (Dec. 5, 1958):14-16. illus., map.
         Describes China's plan to increase petroleum production
      with Soviet assistance by stepping up exploration, prospect-
      ing, and drilling.

C21   "Communist China's Oil Industry." Far Eastern Economic Review
      26, no. 14 (Apr. 2, 1959):470-71. maps.
         Evaluates the previous achievements and future potential
      of China's oil industry, which boosted oil output in 1959 to
      1.5 million tons from a very low base in 1949.

C22   "Petroleum Industry in Taiwan," by K.Y. King. Industry of
      Free China, 12 (Sept. 1959):15-21. illus.
         An overall review of the operation and performance of the
      natural gas and petroleum industry in Taiwan. Covers explora-
      tion, drilling, production, and refining.

C23   "Built with Soviet Aid--The Lanchow Oil Refinery," by Ling-lin Tang. China Reconstructs, 8 (Nov. 1959):12-14. illus.
     Describes the first phase of construction at the Lanchou oil refinery in 1958. Soviet specialists advised and trained Chinese workers in the building process. Soviet materials and equipment accounted for 85% of the sixteen-unit facilities, including a catalytic cracking unit for making aviation fuel and a propane deasphalting unit for producing lubricants.

C24   "New Discovery of Petroleum in Taiwan," by K.Y. King. Industry of Free China 13, no. 3 (Mar. 1960):9-12.
     The author is optimistic about the prospect of oil and gas discovery in Taiwan, despite the cost of exploration. He maintains that geological characteristics of Taiwan's west coast warrant substantial investment in hunting for gas and oil.

C25   "Red China Claims Large Oil Resources Being Developed," by Robert Westgate. World Oil 151 (Nov. 1960):138+. illus., maps.
     Reports on the phenomenal growth in China's oil production between 1950 and 1960 and the increase in exploration in various areas of of the country's interior. Also reports on Soviet and Hungarian assistance in the expansion and building of the Lanchou and Nanking oil refineries, respectively, and the intensified prospecting and geological survey conducted in the Sungliao basin of Manchuria.

C26   "Petroleum and Electric Power Industries," by Chang-yen Li. In Current China, 1949-1959, edited by the Union Research Institute. Hong Kong: Union Research Institute 1 (1961): 177-212. Translation of Shih-nien lai-ti Chung-kang shih-yu ho tien-li kung-yeh.
     An unbiased survey and analysis of China's oil and power industries from the inception of Communist government in 1949 to 1959.

C27   "Petroleum Industry, 1958-1961," by Ladislao La Dany. China News Analysis 406 (Feb. 2, 1962):1-7.
     A brief description of the early stage of the development of China's oil industry.

C28   "China's Shift of Emphasis: The Government Aims at Self-Sufficiency in Oil by 1967, Which Will Be No Easy Goal to Reach." Petroleum Press Service 29 (Nov. 1962):408-10. map, table.
     Discusses China's plans for boosting oil production by 1967. Also examines the current state of the country's oil industry.

C29   "A Study of the Geology and Petroleum Potentialities of
Paoshan and Chingtsaohu Area, Hsinchu," by C.T. Chung.
Petroleum Geology of Taiwan 2 (1963):221-38. illus.
     Reports on the results of a geological survey conducted
around the Chingtsaohu-Paoshan area, where wildcat drilling
was to be executed in later years.

C30   "Petroleum Power of China," by Shri Sailen Gosh. All India
Congress Committee Economic Review 14, nos. 16-17 (Jan. 26,
1963):125-31.
     A critical appraisal of the state of China's petroleum
industry and resources in 1962. Drawing information from
Petroleum Press Service and Petroleum Times, Gosh reviews
China's efforts in oil exploration, estimates her proven oil
deposits, and evaluates China's production and supply of synthetic fuel, her refining capacity, and the need for imported
oil from the USSR and Rumania.

C31   "Petroleum Resources and Production in Mainland China," by
Kuei-sheng Chang. Analysis of Current Chinese Communist
Problems (June 1963):1-30. map, tables.
     One of the most comprehensive reviews of the development
of China's petroleum industry from 1950 to 1962, including
her oil resources (with a breakdown on fields and refineries),
synthetic fuels from oil shale, and an analysis of problems
and weaknesses.

C32   "The Era When We Depend on 'Foreign Oil' Is Gone for Good," by
Ssu-ma t'se. Selections from China Mainland Magazines 406
(Mar. 2, 1964):11-12. Translated from Chung-kuo Ch'ing-nien
(China's Youth) (Jan. 16, 1964).
     Claiming that China has achieved basic self-sufficiency in
oil and petroleum products by the end of her second five-year
plan (1958-1962), the author states that the era in which
China depended on imported oil (from the USSR) is "gone
forever."

C33   "Oil--China's Gap," by G.E. Pearson. Far Eastern Economic
Review 43, no. 5 (Jan. 30, 1964):275-76.
     The author doubts that China was self-sufficient in
petroleum products in 1963, since there was a sizable gap
between her production and consumption.

C34   "Recent Development in the Petroleum Industry." Union
Research Service 35, no. 5 (Apr. 17, 1964):62-80.
     Contains translations of articles on the petroleum indus-
try in China that appeared from January to April 1964 in the
following Chinese newspapers: Yan-ch'eng Wan-pao (Kwangchou),
Jen-min Jih-pao (Peking), Wen-hui Pao (Hong Kong), Hong Kong
Tiger Standard, and Hsin-min Wan-pao (Shanghai).

C35   "From Dependence on 'Foreign Oil' to Basic Self-Sufficiency in
       Petroleum." Selections from China Mainland Magazines, no. 419
       (June 1, 1964):13-14. Translated from Shih-shih Shou-t'se
       (Current Events) (Apr. 21, 1964).
            Underscores China's success in achieving basic self-
       sufficiency in the supply of petroleum and its products and
       describes her efforts to build an integrated petroleum
       industry.

C36   "Has China Enough Oil?" by Daniel Tretiak. Far Eastern Eco-
       nomic Review 44, no. 23 (June 11, 1964):536-37.
            Analyzes China's claim to self-sufficiency in crude oil
       production. Chinese efforts to increase petroleum output
       has resulted in substantially reduced imports from the Soviet
       Union and Rumania. China seems to be producing 7-8 million
       tons of crude a year, while her annual needs are 8-10 million
       tons. Accordingly, 1963 has marked a turning point in China's
       quest for self-sufficiency in oil production.

C37   "Petroleum in China," by Sailen Ghosh. India Quarterly 20,
       no. 3 (July/Sept. 1964):258-80.
            Deals with a broad spectrum of the oil industry in China,
       including China's claim to self-sufficiency in oil production,
       estimates of proved reserves, and highlights of exploration,
       oil field development, refining industries, synthetic oil, oil
       imports, and discoveries of new fields. Most sources are
       drawn from books or articles in the U.S. and China.

C38   "Petroleum Industry in the Chinese Mainland," by Ke-chung
       Chen. Chinese Communist Affairs 1, no. 3 (Sept. 1964):9-18.
       map.
            Chen reviews and assesses the achievements of China's
       petroleum industry from 1950 to 1964 in the context of its
       organization, and the consequences of Soviet assistance. He
       also evaluates China's major oil fields, key oil refineries,
       synthetic oil plants and oil shale deposits, petroleum re-
       sources, demand and supply of crude oil, and the general out-
       look for the industry.

C39   "China Is Basically Self-Sufficient in Oil," by Shih Yen.
       Peking Review 7, no. 42 (Oct. 16, 1964):19-21. illus.
            Stresses the magnitude of China's successful self-
       reliance in building her own petroleum industry and describes
       new achievements in meeting her domestic need for petroleum.

C40   "On the Geology of the Cenozoic Geosyncline in Middle and
       Northern Taiwan (China) and Its Petroleum Potentialities," by
       A. Schreiber. Petroleum Geology of Taiwan, 4 (1965):26-81.
       illus.
            A study on the sedimentary structure of petroleum geology
       in Hsinchu and central Taiwan.

C41   "China's Response: The Chinese, without Foreign Assistance,
Now Manage to Produce the Minimum Amount of Oil Required for
Their Own Economy. . . . Petroleum Press Service 32 (Mar.
1965):88-90.
       Reports that China appears to have overcome serious diffi-
culties since the withdrawal of Soviet experts in 1960. Crude
oil output rose from 200,000 tons in 1950 to 6.6 million tons
in 1960. The 1964 figure is estimated at 8.5 million tons.
The Soviet pullout halted the building of a 3-million-ton oil
refinery in Nanking. Slow progress in oil refining in China
might result in a crude oil surplus, which could be exported
to Japan.

C42   "China's Petroleum Industry--(1)," by Brian Heenan. Far
Eastern Economic Review 49, no. 13 (Sept. 23, 1965):565-67.
illus., map.
       The first of a two-part series on the early development of
China's petroleum industry. Describes how the industry has
progressed, with Soviet help, from its primitive condition to
its current relative self-sufficiency in both plant and equipment. Outlines Soviet contributions in machinery, prospecting
tools, and experts. He also reports that China has recovered
from earlier difficulties caused by the withdrawal of Soviet
aid and has established a machine-building industry sophisticated enough to manufacture most of the plant and equipment
needed for her emerging petroleum industry.

C43   "China's Petroleum Industry--(II)," by Brian Heenan. Far
Eastern Economic Review 50, no. 6 (Oct. 14, 1965):93-95.
table.
       Analyzes China's production of crude oil from 1949 to 1964
and evaluates the historical development of her refining industry. Heenan attributes the sharp decline in oil imports
from the Soviet bloc to the expansion of China's own oil output. He also discusses the present range of petroleum products manufactured in China and the prospects of China's becoming an oil exporting country.

C44   "The Taching Oil Field Develops under the Direction of the
Thought of Mao Tse-tung," by Yu Min. Selections from China
Mainland Magazines 505 (Jan. 3, 1966):33-40. Translated from
Hungch'i, no. 13, (Dec. 6, 1965).
       Describes the achievements and progress in the development
of the Tach'ing oil base.

C45   "How Much Oil?" Economist 218 (Jan. 8, 1966):96-97.
       A penetrating editorial on China's claim that it was
basically self-sufficient in petroleum needs in 1963.

C46   "An Outstanding Oilfield: The Taching Example," by Hsin-yu
       Jen. Peking Review 9, no. 19 (May 6, 1966):33-35. illus.
       Reports on important gains in the production of crude oil
       at the Tach'ing oil field during 1964 and 1965.

C47   "Lan-chou Oil Refinery: An Example of Chinese Road to Indus-
       trialization." Hsinhua Selected News Items (Hong Kong),
       nos. 33-34 (Oct. 3, 1966):31-33. illus.
       Originally featured in a Hsinhua News Agency release on
       September 23, 1966. An account of how workers at the refinery
       succeeded in transforming the Soviet-built facilities into
       technologically advanced models capable of dewaxing and pro-
       ducing hundreds of varieties of high-grade petrochemical
       products.

C48   "Manage Well Oil-fields for the People," by Falien Hu.
       Selections from China Mainland Magazines, no. 548 (Oct. 31,
       1966):11-16. From Hungch'i (Red Flag) (Oct. 1, 1966).
       A young woman worker at Tach'ing oil field tells her own
       story of becoming a model oil field worker by persistently
       studying Mao's thought and putting it into practice to serve
       the people.

C49   "Chinese Enigma." Petroleum Press Service 33, no. 11 (Nov.
       1966):406-7.
       Reports that the spirit of the Tach'ing oil field, repre-
       senting China's success in self-reliance, is being used as a
       model to revolutionalize the whole economic and social life of
       700 million Chinese.

C50   "Technology in China," by Genko Uchida. Scientific American,
       215, no. 5 (Nov. 1966):37-45. illus., tables.
       An overall appraisal of the state of China's industrial
       technology in the 1960s, including her chemical and petro-
       chemical industries.

C51   "Fill'er up." Newsweek, 68 (Nov. 21, 1966):71. illus.
       Reports that China's oil industry has made large strides
       since the withdrawal of 500 Soviet oil specialists in 1960.
       The Tach'ing oil field, initially developed with the help of
       Soviet technological knowhow in 1959, has become the country's
       most productive oil base. Two additional oil fields, Shengli
       and "Industry no. 1009," have been developed, and crude oil
       production in 1966 is estimated at 10 million tons.

C52   "To Develop an Oil Field, Get Rid of Ghosts and Ogres." Oil
       and Gas Journal 64, no. 48 (Nov. 28, 1966):126, 130-31. illus.
       A profile of the Tach'ing oil field in its earlier phase.
       Describes the tremendous effort and investment made, from its
       discovery in 1960 to its substantial oil production in 1965.
       The Tach'ing oil field is estimated to produce more than one-
       third of China's estimated 180,000 b/d of crude oil in 1965.

C53   "Guided by Mao Tse-tung's Thought, China Introduces Fermenta-
       tion Dewaxing Process in Oil." Peking Review 9, no. 50
       (Dec. 9, 1966):32-34.
           A new technology for removing paraffin content from crude
       oil during the refining process has been developed by techni-
       cians in Shanghai.

C54   "Chinese Red Claims Catalytic Reformer Built from Scratch."
       Oil and Gas Journal 65, no. 4 (Jan. 23, 1967):62. illus.
           Deriving its information from China Reconstructs, the
       Journal states that China has succeeded in building a new
       catalytic reforming plant using its own technology and tech-
       nicians. The plant extracts high-purity benzene from petro-
       leum, and makes raw materials for insecticides, synthetic
       fibers, dyestuffs, and other byproducts. The plant was de-
       signed and built completely by the Chinese.

C55   "The Chinese Communist Petroleum Industry," by Paul L.C. Hao.
       Chinese Communist Affairs, 4, no. 1 (Feb. 1967):20-26.
           Primarily updates and upgrades the similar article "Petro-
       leum Industry in the Chinese Mainland," by Ke-chung Chen
       (entry C38), which appeared in the September 1964 issue of
       this journal.

C56   "Communist China and Petroleum," by Jan S. Prybyla. Military
       Review, 47, no. 2 (Feb. 1967):48-53. illus.
           An updating of an earlier article, published by the author
       in the February 1965 issue of the Review (45:19-22). Provides
       a general survey and analysis of China's petroleum industry--
       oil reserves, production, and refining. Also discusses the
       statistical problem of quantifying China's oil production and
       verifying her claims to major gains--a problem occasioned by
       the unavailability of hard statistical figures.

C57   "Internal Strife May Hurt Communist Chinese Oil Effort." Oil
       and Gas Journal 65, no. 8 (Feb. 20, 1967):36.
           According to Chinese press and radio reports, oil produc-
       tion and refining have been affected by internal strife in the
       Sinkiang and Heilungkiang areas. A 60,000 b/d oil refinery
       located in Sinkiang is reportedly to be the center of bloody
       battles. Reports that work at the Tach'ing oil field has
       halted because 10,000 student-workers have left for Peking to
       participate in Red Guard activities.

C58   "Red China Battles for Oil Status." Oil and Gas Journal 65,
       no. 8 (Feb. 20, 1967):35-36. illus., maps.
           Discusses the secrecy of China's oil activity and the
       vital role the Soviets played in assisting China in oil ex-
       ploration, drilling, production, refining, and the building of
       plant and equipment. The Soviets helped train China's oil
       experts, build an oil refinery at Lanchou, and provide seismic

and aerial magnetic techniques for geological surveys. Despite the suspension of Soviet aid, China has obtained some technological help from Rumania and other East European countries, succeeded in developing the Tach'ing oil field, and stepped up drilling activity in various provinces, such as the promising Tarim and Tsaidam basins.

C59 "Soviets Hint Red China's Taching is a Paper Tiger." Oil and Gas Journal 65 (June 26, 1967):57.
Reports the USSR's downgrading of the Tach'ing oil field's importance. The official Soviet party paper Izvestia cast doubt on China's claim to self-sufficiency in oil production, which is largely attributed to Tach'ing's output of crude oil.

C60 "China Keeps the Oil Flowing--The World at Work," by Colina MacDougall. Times Review of Industry and Technology 5, no. 8 (Aug. 1967):58-59. map.
Reports that despite the disruption caused by the Great Cultural Revolution, China seems to be able to keep a steady flow of crude oil from her major fields. Apparently, Chinese authorities are doing their best to protect the oil industry from the destructive effects of political turmoil.

C61 "A 'New Phoenix' Rises from a Nest of Straw." China Pictorial 9 (Sept. 1967):50-52. illus.
Reports on the Harbin Petroleum Machinery Accessories Plant.

C62 "Photogeological Observations on the Low Hilly Terrain and Coastal Plain Area of Hsinchu, Taiwan," by C.H. Tang. Petroleum Geology of Taiwan 6 (1968):71-79. illus.
A geological report on the hydrocarbon potential of the coastal area of northern Taiwan.

C63 "A Stratigraphic and Sedimentary Analysis of the Protoquartzite in the Miocene Talu Shale in Northern Taiwan," by J.T. Chou. Petroleum Geology of Taiwan 6 (1968):115-36. illus.
A geological study on the extension of the continental shelf in northern Taiwan, which is presumed to possess hydrocarbon potential.

C64 "Natural Gas and Opportunities for Taiwan," by Frank I. Baxter. Industry of Free China 29, no. 2 (Feb. 1966):20-29.
Foresees potential benefits for Taiwan in the development and utilization of her natural gas resources and urges the formation of a national fuel policy patterned after that of the U.S. for industrial and residential allocation and distribution.

C65   "Turmoil Cripples Red Chinese Oil." <u>Oil and Gas Journal</u> 66,
       no. 22 (May 27, 1968):73-74.
            States that the Cultural Revolution in 1967 caused turmoil
       in oil production and refining centers, resulting in severe
       losses. According to Japanese and Soviet sources, damages
       have been greater than during the period of the Great Leap
       Forward. Reports are circulating that production, refining,
       and transportation have been most affected by the turmoil,
       while exploration has not been greatly harmed.

C66   "Geologic Concepts Relating to the Petroleum Prospects of
       Taiwan Strait," by Meng Chao-yi. U.N. Economic and Social
       Commission for Asia and the Pacific, Committee for Coordina-
       tion of Joint Prospecting for Mineral Resources in Asian
       Offshore Areas. <u>Technical Bulletin</u> 1 (June 1968):143-53.
       illus., maps.
            Discusses Taiwan's offshore oil potential in the Straits
       of Taiwan, which is part of China's outer continental shelf.

C67   "Black Magic," by Ming Liang. <u>Free China Review</u> 18, no. 7
       (July 1968):11-15. illus.
            Reports that the opening of naphtha and polyethylene
       plants in the port city of Kaohsiung heralds a new era in
       which the petrochemical industry in Taiwan will make a sig-
       nificant contribution to the island's economy.

C68   "Peiping's Petroleum Industry," by Ko-jen Ho. <u>Issues &</u>
       <u>Studies</u> 4, no. 11 (Aug. 1968):22-35. tables.
            Summarizes China's petroleum industry in three stages:
       1953-1957, 1958-1962, and 1963-1967. This broad survey of
       China's oil industry covers reserves, production, refining,
       storage, transportation, and petroleum products--a rather
       unbiased assessment from Taiwanese sources. It provides
       seven tables to indicate the relative position of China's
       oil industry.

C69   "Taiwan's Big Push: Petrochemicals." <u>Chemical Week</u> 103
       (Aug. 24, 1968):60-62. illus.
            Reports that Taiwan has given its petrochemical industry
       top priority in its fourth four-year plan. Taking advantage
       of natural gas produced in the north, the China Petroleum
       Corp. supplies ample materials for downstream operations.
       U.S. companies are heavily involved in these developments.
       Companies such as Allied Chemical, Atlas Chemical, Gulf Oil,
       Mobil Chemical, and National Distillers invested $128 million
       in 95 projects, mostly in the petrochemical industry, produc-
       ing a variety of products, such as PVC, urea fertilizer,
       resins, ethylene, and propylene.

C70 "China's Taching Oilfield: Eclipse of an Industrial Model."
Current Scene 6, no. 16 (Sept. 17, 1968):1-10.
Analyzes the possible downgrading of the Tach'ing oil
field, China's prime industrial model. Attributes Tach'ing's
fall from grace during the height of the Cultural Revolution
in 1967 and 1968 to the awareness by China's political radi-
cals that Tach'ing's success is largely due to heavy state
investment and classic capitalistic management rather than to
pure political and ideological motivation.

C71 "Oil and Euphoria--Japan-China Trade." Economist 268
(Sept. 23, 1968):96-97.
Reports that Japan's hoped-for mushrooming of trade with
China hinges on a big increase in the export of Chinese crude
oil to Japan. The Japanese estimate that by holding down her
consumption of oil, China can export up to 1 million b/d of
crude oil to Japan by 1983 or 1984, thus earning about $5 bil-
lion a year to repay loans for Japanese industrial equipment.

C72 "Excellent Situation in Taching Oilfield." Peking Review 11,
no. 44 (Nov. 1, 1968):29-30. illus.
A news release on the formation of the Tach'ing Oil Field
Revolutionary Committee, reporting that state goals were surpassed for crude oil production and processing. Daily crude
oil production has topped its previous peak and oil field
construction scheduled for the third five-year plan has been
completed in three years.

C73 "Taiwan Ethylene Plant On-Stream." Oil and Gas Journal 66,
no. 45 (Nov. 4, 1968):129. illus.
A new naphtha-cracking facility in Kaohsiung (southern
Taiwan), has been completed, with a production capacity of
120 million pounds of ethylene.

C74 "Notes on the Taching Oilfield." China Reconstructs 17
(Dec. 1968):38-43. illus. (some color).
An account of a model oil drill team (no. 1205) in action,
as well as a story on a coking plant in the Tach'ing oil
refinery.

C75 "China's Oil Hurt by Revolution." Oil and Gas Journal 66,
no. 50 (Dec. 9, 1968):144-49. illus.
The Soviets estimate that China's oil production declined
to 200,000 b/d in 1967, from 240,000 b/d in 1966. One of the
casualties of the Cultural Revolution has been the disruption
of operations in China's largest petrochemical complex, the
Tach'ing oil field. Disruptions in the metallurgical and
power industries have had a serious impact on China's oil
production and refining and reportedly have caused a large
number of workers to flee Tach'ing.

C76 "China's First Deep-Well Drilling Machine." <u>Peking Review</u> 11, no. 51 (Dec. 20, 1968):29. illus.

Reports that China's first deep-well rig has been built in only six months by the Lanchou Petrochemical Machinery Works. This Chinese-designed and built machine has passed its trial test by drilling through complex geological strata and several high pressured oil-bearing formations. The rig weighs 300 tons and is powered by a 3,000-horsepower diesel engine.

C77 "Red China's 'Lost' Oil Field Is Found." <u>World Oil</u> 166, no. 1 (Jan. 1969):9.

The mystery surrounding the exact location of China's Tach'ing oil field has been solved with the disclosure by Taiwan authorities that the site is located in Manchuria, eighty miles northwest of Harbin.

C78 "Taching Workers Courageously Stride Forward." <u>China Pictorial</u> 2 (Feb. 1969):26-31. illus. (some color).

Extensive photo coverage of the Tach'ing oil field two years after the advent of the Cultural Revolution. Despite the upheaval caused by the Cultural Revolution, the Tach'ing oil base has forged ahead with the establishment of its Revolutionary Committee (on May 31, 1968), and the construction tasks were completed three years ahead of schedule.

C79 "Red China Reports Higher Production." <u>Oil and Gas Journal</u> 67, no. 5 (Feb. 3, 1969):126-27.

Discusses reports from China on the substantial gains made in oil production during 1968. The <u>Journal</u>'s estimate of China's crude oil output is 116,000 b/d in 1962 and 240,000 b/d in 1966, with a possible increase to 260,000 b/d in 1969. It is believed that China's recoverable oil reserve is in the vicinity of 15 billion barrels.

C80 "First Big Oil Tanker Successfully Launched--Good News from China's Industrial Front." <u>Peking Review</u> 12, no. 16 (Apr. 18, 1969):35-36. illus.

Reports on the construction of China's first 15,000-ton oil tanker at the Hungch'i shipyard in Talien. Built with Chinese-made steel, <u>Tach'ing No. 27</u> has been successfully launched in a short period of time, with its hull completed in only 32 working days.

C81 "Taching Oil Hikes Red Chinese Output." <u>Oil and Gas Journal</u> 67, no. 16 (Apr. 21, 1969):62-63. illus.

Estimates 1968 crude oil output at Tach'ing at 50,000 b/d, a 34% increase over 1967.

C82 "Geological Structure and Some Water Characteristics of the
East China Sea and the Yellow Sea," by K.O. Emery et al.
U.N. Economic and Social Commission for Asia and the Pacific,
Committee for Coordination of Joint Prospecting for Mineral
Resources in Asian Offshore Areas. Technical Bulletin 2
(May 1969):3-43. illus., maps.
    Probably one of the most significant reports on the tec-
tonic survey of the sediments of the East and South China
seas. The survey was conducted by research vessel Hunt under
the sponsorship of the United Nations. The results of this
survey have raised many speculations about the size of hydrocarbon deposits in China's continental shelf. This U.N. report indicates that the magnitude of China's offshore petroleum resources could be comparable to or even greater than
many oil fields in the Persian Gulf. (See also entry C100.)

C83 "Reports on the Seismic Refraction Survey on Land in the
Western Part of Taiwan, Republic of China," by Y.K. Sato et
al. U.N. Economic and Social Commission for Asia and the
Pacific, Committee for Coordination of Joint Prospecting for
Mineral Resources in Asian Offshore Areas. Technical Bulletin
2 (May 1969):45-57. illus., maps.
    This seismic survey of Taiwan's western coastal area re-
ported rich potential reserves of oil and gas.

C84 "15,000-Ton Oil Tanker Launched in China." Economic Reporter
(English supp.) 2 (Apr./June 1969):34. illus.
    Describes the successful launching of China's first
15,000-ton tanker designed and constructed completely by the
Chinese. It took only 32 days at Hungch'i (Red Flag) shipyard
in Talien to build the tanker, from the laying of its keel to
launching of the ship.

C85 "China's Petroleum Industry Advances with Big Strides on Road
of Self-reliance." Peking Review 12, no. 39 (Sept. 26, 1969):
17-19.
    Reports that China has established a modern oil industry.
From the standpoint of quality, quantity, and variety, more-
over, China is now self-sufficient in the petroleum products
needed for her national economy and defense. Indeed, some
aspects of her oil technology have already reached or have
surpassed advanced world levels.

C86 "Red China Claims Oil Self-sufficiency." Oil and Gas Journal
67 (Oct. 13, 1969):62.
    Reports on China's claim to basic self-sufficiency in the
domestic supply of crude oil. While no statistical figures
have been given on the production of crude oil from 1959,
Western sources estimate China's oil output in 1965 to ap-
proach 10 million tons (203,000 b/d). Also reports are cir-
culating on the possible discovery of offshore oil in the
Pohai Gulf.

C87   "China's Petroleum Industry," by Sien-chong Niu. Military
       Review 49, no. 11 (Nov. 1969):23-27. illus.
            A general survey and assessment of China's petroleum in-
       dustry, with emphasis on the estimation of her oil resources.
       Although China claims to be basically self-sufficient in
       petroleum and petroleum products, her inadequate supply of oil
       continues to restrain her from venturing into war, even with
       the intensification of the Vietnam conflict near her southern
       border.

C88   "New Forward Leap in China." Petroleum Press Service 36
       (Nov. 1969):405-7. map, table.
            Analyzes China's claim to self-sufficiency in producing
       an adequate quantity, variety, and quality of petroleum prod-
       ucts. The Service examines production figures for 1960-1962,
       1963-1966, and 1967-1969, focusing on some production set-
       backs, bottlenecks in technology, and the importance of
       Tach'ing to overall crude oil production in China. Despite
       the big increase in crude oil output claimed by China, fre-
       quent disruptions of oil output have been caused by internal
       strife during the Cultural Revolution.

C89   "Basic Chemical and Petrochemical Industry in the Republic of
       China," by W.K. Chen. Economic Review (International Commer-
       cial Bank of China, Taipei) 132 (Nov./Dec. 1969):12-15.
            Summarizes the development and progress of Taiwan's petro-
       chemical and chemical industry, its problems, and its poten-
       tial for growth.

C90   "China Petroleum Industry in Soaring Progress." Asian Outlook
       5, no. 1 (Jan. 1970):26.
            Describes the growth and expansion of the China Petroleum
       Corp. of Taiwan in the past twenty years. The corporation is
       currently engaged in the exploration, refining, marketing, and
       transportation of petroleum and its products in the Taiwan
       area.

C91   "Taching Oilfield Wins Tremendous Success in Revolution and
       Production." Peking Review 13, no. 2 (Jan. 9, 1970):29-30.
            A news summary on the success at the Tach'ing oil field
       during the year 1969. Compared with 1968, Tach'ing has made
       the following gains: the production capacity of crude oil was
       up 40.7%; the total industrial output value rose by 21%; and
       oil field construction increased by 10%.

C92   "Red China Reports Taching Output Up." Oil and Gas Journal
       68 (Feb. 9, 1970):27.
            Reports on Peking's claim of big gains in the production
       of crude oil (a 41% increase over the previous year) and the
       construction of oil fields at Tach'ing in 1969. Also reports
       on Soviet doubts about China's achievement, as carried by Tass.

C93   "Preliminary Study of Submarine Geology of China's East Sea and the Southern Yellow Sea," by Yun-shan Ch'in and Shih-ch'ing Fan. Translations on Communist China, no. 97 (Apr. 7, 1970):12-36. (Joint Publications Research Service 50252.) Translation from the original in Hai-yang yu hu-chao (1959).

    Deals with the tectonic study of submarine sediments in China's continental shelf for possible offshore oil deposits.

C94   "Airomagnetic Survey of Offshore Taiwan," by W. Bosum et al. U.N. Economic and Social Commission for Asia and the Pacific, Committee for Coordination of Joint Prospecting for Mineral Resources in Asian Offshore Areas. Technical Bulletin 3 (May 1970):1-34. illus.

    Reports on a geological magnetic survey of Taiwan's offshore waters, particularly the area of the Strait of Taiwan.

C95   "Big Advances in the Oil Industry," by Shang-yu Shi. China Reconstructs 19, no. 5 (May 1970):2-4. illus.

    Summarizes the growth and development of China's petroleum industry from its inception in 1950 to the present. Stresses China's self-sufficiency in all petroleum products in terms of their quality and quantity. Reports that China can match advanced world levels in many areas of oil technology.

C96   "A Conception of the Evolution of the Island of Taiwan and its Bearing on the Development of the Neogene Sedimentary Basins on its Western Side," by Chao-yi Meng. U.N. Economic and Social Commission for Asia and the Pacific, Committee for Coordination of Joint Prospecting for Mineral Resources in Asian Offshore Areas. Technical Bulletin 3 (May 1970):109-26. illus.

    A study of the tectonic structure of Taiwan's west coast facing the Strait of Taiwan.

C97   "Note on Sea Bottom Sampling in Offshore Area of Taiwan, China," by Chinese Petroleum Corp. U.N. Economic and Social Commission for Asia and the Pacific, Committee for Coordination of Joint Prospecting for Mineral Resources in Asian Offshore Areas. Technical Bulletin 3 (May 1970):35-36. illus.

    Reports on the examination of sea bottom rocks in the Strait of Taiwan as part of a geological survey of Taiwan's offshore oil potential.

C98   "China Nurses Crude Flow Back Up to '66 Level." Oil and Gas Journal 68, no. 26 (June 29, 1970):28-29. illus.

    Estimates China's crude oil output in 1969 at 240-260,000 b/d, in contrast to Soviet analysts' downgrading of China's performance in the same year.

C99　　"Development in Mainland China, 1949-1968," by A.A. Meyerhoff.
　　　　<u>Bulletin of the American Association of Petroleum Geologists</u>
　　　　8 (Aug. 1970):1567-80.　maps.
　　　　　　One of the most authoritative assessments of China's
　　　　petroleum industry from 1949 to 1968.　The author traces the
　　　　history of China's exploration from 211 B.C. to 1949 and de-
　　　　scribes the stages of petroleum exploration from 1949 to 1968.
　　　　He identifies 16 major oil-bearing basins in the tectonic
　　　　geology of China, and provides data on 54 key producing
　　　　fields, including 38 oil fields and 16 gas fields.　China's
　　　　oil reserves, production estimates, pipelines, and refineries
　　　　are discussed in detail.　Meyerhoff credits the Communist
　　　　Chinese authorities with the discovery of 48 oil fields and
　　　　the expansion of 13 existing fields.　Total proven and proba-
　　　　ble reserves are estimated at 19.6 billion barrels, and annual
　　　　production of crude oil is believed to have reached 100 mil-
　　　　lion barrels.　He considers China to be basically self-
　　　　sufficient in oil, but cites the shortage of pipelines and
　　　　refining equipment, and inadequacies in related logistical
　　　　support as factors retarding the future development of her
　　　　oil industry.

C100　　"Structural Framework of East China Sea and Yellow Sea," by
　　　　John M. Wageman, Thomas W.C. Hilde, and K.O. Emery.　<u>Bulletin
　　　　of the American Association of Petroleum Geologists</u> 9 (Sept.
　　　　1970):1611-43.　maps.
　　　　　　Reports on the results of the geophysical survey in the
　　　　East China Sea and the Yellow Sea conducted by scientists
　　　　from the member nations of the U.N. Economic Commission for
　　　　Asia and the Far East aboard the research ship <u>Hunt</u> during
　　　　October and November 1968 (see also entry C82).　Sedimentary
　　　　strata with possibly vast hydrocarbon potential were found
　　　　beneath the continental shelf.　This is the first of the two
　　　　most important oil and gas surveys ever conducted in China's
　　　　offshore areas.

C101　　"Red China Near Self-sufficient in Oil."　<u>Oil and Gas Journal</u>
　　　　68, no. 40 (Oct. 5, 1970):78-79.
　　　　　　The <u>Journal</u>'s speculation on the claim that China is
　　　　almost self-sufficient in the supply of oil.　A.S. Meyerhoff,
　　　　an authoritative geologist from the American Association of
　　　　Petroleum Geologists, believes that the crude oil output in
　　　　China could hit 1 million b/d by the end of 1969, thus ef-
　　　　fectively ending her need to import oil from other communist
　　　　countries.

C102　　"China's Petroleum Industry Develops with Greater, Faster,
　　　　Better, and More Economical Results."　<u>Peking Review</u> 13,
　　　　no. 41 (Oct. 9, 1970):30-31, 37.
　　　　　　Reports and comments on the speed with which the petroleum
　　　　industry in China is being built.　Also reports that China's

oil industry exceeded its 1970 production target by a wide margin. Crude oil output for the period of January to August 1970 rose 34% from the corresponding period in 1969. The production of key refined products such as gasoline, kerosene, diesel oil, lubricants, fuel oil, and asphalt hit an all-time high.

C103   "Petrochemicals Getting Top Taiwan Priority." Chemical Week 107 (Oct. 21, 1970):77-78. map.
     Reports that the petrochemical industry has been given first priority in Taiwan's economic planning and will receive an outlay of $200 million for the building of new plants and the expansion of existing ones. This includes constructing an ethylene plant with a capacity of 600 million pounds a year, doubling the capacity for polyethylene production, and doubling the capacity of the Kaohsiung oil refinery--100,000 b/d by 1971. At the moment many U.S., German, Italian, and Japanese firms are keenly interested. The plan will be financed through the Export-Import Bank and the International Finance Co.

C104   "China's Crude Output Rises to Meet Needs." World Petroleum 41, no. 11 (Nov. 1970):16.
     Comments on the report by the Bulletin of the Association of American Petroleum Geologists that China had become nearly self-sufficient in the production of crude oil (see entry C99).

C105   "Great Victories for China's Petroleum Industry." Economic Reporter (English supp.) 4 (Oct./Dec. 1970):22-23. illus.
     Summarizes the oil industry's achievements. Emphasizes the role played by China's largest oil base, the Tach'ing oil field.

C106   "Mao-Tse-tung Thought Brings New Life to Northwest China Oilfield." Economic Reporter (English supp.) 4 (Oct./Dec. 1970): 31-32. illus.
     Profiles the development of one of China's oldest oil fields, Yümen oil field in Kansu province. Reports that great progress has been made in the past five years for this old oil base in northwest China. Crude oil output has more than doubled, from wells once considered depleted.

C107   "Genzoic Basin of Western Taiwan--Case History of the Chingtsaohu Gas Field, Taiwan, China," by Stanley Chang et al. In Case Histories of Oil and Gas Fields in Asia and the Far East. New York: U.N. Economic Commission for Asia and the Far East, 1971, pp. 43-57. illus., maps.
     Describes the exploration and development of oil and gas at the CT-1 wildcat well in the southern Hsinchu area in central Taiwan.

C108   "The Taching Oil Field." In <u>Collected Documents of the First</u>
        <u>Sino-American Conference on Mainland China</u>. Taipei:
        Institute of International Relations, Republic of China,
        1971, pp. 795-820.
            An evaluation of China's leading oil field, Tach'ing, made
        by international scholars at a Taipei conference about China.

C109   "Chinese Shipyard Builds 10,000-Ton Class Oil Tanker."
        <u>Economic Reporter</u> (English supp.) 1 (Jan./Mar. 1971):24-25.
        illus.
            Reports that the first completely Chinese designed and
        built oil tanker in the 10,000-ton class was launched at the
        Hungch'i (Red Flag) shipyard in Talien. The tanker was named
        <u>Tach'ing No. 29</u>.

C110   "Taching Oilfield, Pt. I: Race against Time," by Anna Louise
        Strong. <u>China Now</u>, no. 11 (Apr./May 1971):8-9.
            Describes the speed with which people and materials were
        mobilized to build the Tach'ing oil base.

C111   "Country Report: China (Taiwan)," by Eoin H. MacDonald. U.N.
        Economic and Social Commission for Asia and the Pacific, Com-
        mittee for Coordination of Joint Prospecting for Mineral Re-
        sources in Asian Offshore Areas. <u>Technical Bulletin</u> 5 (May
        1971):32-47. illus., tables.
            A general review of the prospects and potentials of
        Taiwan's offshore petroleum resources.

C112   "New Oilfield in Chinghai." <u>Peking Review</u> 14, no. 22
        (May 28, 1971):19.
            Reports that a new oil field was discovered in the western
        part of Tsaidam Basin in Tsinghai province, situated at 3,000
        meters above sea level. Initial prospecting and drilling
        started in 1958, but were halted. The exploration resumed in
        April 1969, and a new oil field was found and built within a
        year and a half, together with an oil refinery and a building
        materials plant.

C113   "Taching Red Banner Becomes More Radiant Than Ever." <u>Economic</u>
        <u>Reporter</u> (English supp.) 2 (Apr./June 1971):4-5. illus.
            Describes the achievements and progress made during the
        past ten years in developing the Tach'ing oil field.

C114   "Foraminiferal Trends in the Surface Sediments of Taiwan
        Strait," by Tunyow Huang. U.N. Economic and Social Commission
        for Asia and the Pacific, Committee for Coordination of Joint
        Prospecting for Mineral Resources in Asian Offshore Areas.
        <u>Technical Bulletin</u> 5 (June 1971):23-61. illus., tables.
            Reports on a tectonic study of the surface sediments of
        offshore Taiwan.

C115   "Li Szu-kuang, Prominent Chinese Geologist Dies." <u>Current</u>
       <u>Scene</u> 9, no. 6 (June 7, 1971):13.
            Reports the death of Li Szu-kuang, China's leading scien-
       tist and geologist, at the age of 82. A graduate in geology
       at Birmingham University (England), Li received his Ph.D. in
       1920 and then taught geology at Peking National University.
       A pioneer in his field, Li developed a theory on oil-bearing
       strata in the continental sedimentation and was instrumental
       in the discovery of many oil fields in northeast China and the
       subsequent development of the country's petroleum industry.

C116   "Taiwan Plans $103.6 Million Oil Program." <u>Oil and Gas Jour-</u>
       <u>nal</u> 69 (June 7, 1971):27.
            Taiwan plans to appropriate $103.6 million for oil ex-
       ploration, oil refining, and petrochemical facilities over
       next three years. The fund comes from private investments,
       the government, foreign loans, and the China Petroleum Corp.
       The breakdown of the fund is as follows: $50 million for a
       refinery built at T'aoyüan near Taipei, $38 million for
       transportation (including tankers and pipelines), $9.3 million
       for modernizing the existing plants, and the rest on other
       minor projects.

C117   "Taching Oilfield, Pt. II: The Women," by Anna Louise Strong.
       <u>China Now</u>, no. 13 (July 1971):3-4, 10. illus.
            Strong, a veteran journalist who resided in China most of
       her adult life, describes the role female workers played in
       the building of the Tach'ing oil field, as well as their con-
       tributions in enhancing the community life in the oil field
       and on the farm.

C118   "Gas for Fuel and Industry," by Suyen Chien. <u>Free China</u>
       <u>Review</u> 21, no. 8 (Aug. 1971):25-30. illus., maps.
            Records the exploitation and utilization of Taiwan's
       natural gas reserves in the past twenty years. The steady
       increase in production has enabled the island's booming
       industry to use natural gas as both fuel and raw material for
       manufacturing a wide variety of petrochemical products.

C119   "Taching Oilfield, Pt. III: Death of an Oligopoly," by Mary
       Z. Brittain. <u>China Now</u>, no. 14 (Aug. 1971):3-4. illus.
            Tells of the effect of the successful development of the
       Tach'ing oil field and the adequate supply of oil for China,
       which have ended foreign oil companies' ability to use petro-
       leum as a weapon to gain political and economic leverage
       against China.

C120   "At Taching: Red Banner on Industrial Front." <u>China Pic-</u>
       <u>torial</u> 9 (Sept. 1971):2-12. illus. (some color).
            A profusely illustrated account of the Tach'ing oil field,
       whose rapid construction has made China basically self-

sufficient in petroleum products. The photographs also present the integrated life at Tach'ing, with the men working in
the oil field and the women on the Tach'ing farms.

C121 "China's New Status." Petroleum Press Service 10 (Oct. 1971):
      363–65. map.
            Argues that the visit to China by President Nixon has pro-
      vided China with a new world status and an incentive to take a
      more positive stance toward cooperation with the Western coun-
      tries in developing her oil deposits. The article also cites
      A.A. Meyerhoff's estimate of China's oil and gas production in
      1970 (see entry C99).

C122 "Oil: Statistics and Projections." China Trade Report 10
      (Oct. 1971):11.
            Draws on reports from New China News Agency, Nihon Keizai
      Shimbun (Japan Economic Daily), and Petroleum Press Service
      (London) reports (entry C121), to summarize the estimates and
      projections of China's oil output. Production has probably
      risen from 5.5 million tons in 1960 to an estimated 20 million
      tons in 1970. At present China is more than self-sufficient
      in petroleum products and is ready to export some crude oil to
      Japan.

C123 "5,000-Ton Tanker Built on Sandy Bank." Peking Review 14,
      no. 41 (Oct. 8, 1971):22.
            A small shipyard that had previously constructed only a
      500-ton tanker and a 1,500-ton deck lighter has successfully
      built a 5,000-ton tanker on a sandy bank without a berth. By
      building sections of the vessel and then joining them together,
      the Tsing-tao Hunghsing Shipbuilding and Repairing Plant (in
      Shantung province) has succeeded in completing the Tach'ing
      No. 49.

C124 "Petroleum Industry Success--Progress Report." Peking Review
      14, no. 43 (Oct. 22, 1971):8–9. illus.
            Reports that China increased crude oil output more than
      30% in each year from 1960 to 1970.

C125 "Mainland China--A Message from the Editor." World Petroleum
      42, no. 11 (Nov. 1971):23.
            A comment on the possibility of China's becoming a signif-
      icant power in oil production in view of A.A. Meyerhoff's study
      published in the Bulletin of the Association of American
      Petroleum Geologists (entry C99).

C126 "Chinese Oil Flow Up, but Much Larger Gains Needed," by Frank
      J. Gardner. Oil and Gas Journal 69, no. 50 (Dec. 13, 1971):
      35–39. illus., maps (some color).
            Reports that in 1971, China's oil production recovered
      from the damage caused by the Cultural Revolution in 1966-1968.

Crude oil output could top 25 million tons in 1971, while refining capacity may be raised proportionately. Drilling footage is up 80% from 1970. Oil strikes have been made in the
Sungliao and Tsaidam basins, and there are reports of gas
finds and shallow gas field development in Kweichow, Kwangtung,
and Chekiang provinces.

C127   "Taching No. 30 Oil Tanker Built." Peking Review 14, no. 52
        (Dec. 24, 1971):21.
        A news report on the launching of a 10,000-ton tanker at
        Hungch'i Shipyard in Talien, emphasizing the speed with which
        the tanker was built. The hull of Tach'ing No. 30 was com-
        pleted in only 15 days and installed in only half the time
        needed for a 15,000-ton tanker, Tach'ing No. 29.

C128   "A New Type of Industrial and Mining Area--Report from Taching
        Oilfield." Peking Review 14, no. 52 (Dec. 31, 1971):6-9.
        illus.
        The Tach'ing oil field is characterized by the integration
        of industry and farming, as well as the combination of city
        and country. The article reports that instead of a vast oil
        city where all modern facilities have to be concentrated,
        simple and scattered agri-industrial settlements dot the
        horizon of the Sungliao prairie in the northeast.

C129   "China: A Nation Reaches for the Energy Age," by Susan L.
        Smith. Petroleum Today 13, no. 3 (1972):23-25. illus.
        Even though oil and gas account for about 15% of her total
        energy consumption, China is basically self-sufficient in the
        supply of petroleum products.

C130   "Vigorous Political and Ideological Work--Report from Taching
        Oilfield." Peking Review 15, no. 2 (Jan. 14, 1972):11-12, 19.
        illus.
        A news report on how progress in studying the Maoist doc-
        trine and "learning-from the People's Liberation Army" can
        spur workers' zeal for oil production. This report emphasizes
        the vital role played by ideological education for young
        workers, who when equipped with revolutionary zeal, can often
        overcome physical and technical problems.

C131   "Big Increase in Crude Oil." Peking Review 15, no. 3
        (Jan. 21, 1972):3.
        This brief article reports on gains in crude oil produc-
        tion in 1971. A 28% increase in crude oil output and 25% in
        natural gas production were recorded in 1971 over 1970. Gains
        of 17% and 16% were also registered in China's extracting and
        processing capacity, respectively. New oil deposits are said
        to have been discovered and many new oil fields have been
        opened.

C132   "1205 Team Drills over 127,000 Metres." Peking Review 15,
        no. 4 (Jan. 28, 1972):8-9. illus.
            A profile of Tach'ing's heroic No. 1205 drilling team,
        which has drilled a record 127,000 meters in the year 1971.
        Up to last year, the team already drilled several hundred
        high-yielding oil wells. Back in 1966, it also achieved a
        record footage by drilling 100,300 meters.

C133   "Peking Claims Crude Output Hits 520,000 B/D--Up 28%." Oil
        and Gas Journal 70 (Feb. 7, 1972):25.
            An estimate of China's oil production in 1971, which might
        have reached 26 million tons, a gain of 28% over previous
        year, with consumption estimated at only half the oil pro-
        duced. The crude oil output is expected to increase further
        in 1972 to 30 million tons, and by 1975 it should exceed
        50 million tons a year.

C134   "Industrial Waste Water Serves Agriculture." China Recon-
        structs 21, no. 3 (Mar. 1972):34-35. illus.
            Describes China's success in purifying the polluted water
        released by the Peking General Petrochemical Works and recy-
        cling it for agricultural use. The polluted water is purified
        through chemical treatment and released to local farms for
        irrigation.

C135   "A Petrochemical Works under Construction." China Recon-
        structs 21, no. 3 (Mar. 1972):26-33. illus. (some color).
            Portrays the construction of the Peking General Petro-
        chemical works in progress. Located in southwestern Peking,
        the petrochemical complex has been under construction since
        the autumn of 1968 by more than 10,000 soldiers, technicians,
        and workers. Already, 11 refining units, a synthetic rubber
        plant, and a phenol acetone and polystyrene plant have come
        on stream. The vast complex is being completely designed,
        installed, and built by Chinese with equipment and transporta-
        tion provided by 500 plants in 30 provinces throughout China.

C136   "China Pushing Oil Exploration in Remote Tsaidam Basin," by
        Frank J. Gardner. Oil and Gas Journal 70 (Mar. 27, 1972):
        44-45. illus.
            Reports that Tsaidam basin in Tsinghai province has become
        a focus of oil exploration. Major seismic, wildcatting, and
        development efforts are proceeding within the 77,000-square-
        mile basin. Oil drilling is costly in this 10-20,000-foot-
        high plateau, but the area has a huge petroliferous potential.
        Estimates by American geologists place the Tsaidam area's
        recoverable reserves at 230 million barrels.

C137   "China's New Status in Oil." <u>Survival</u> 14, no. 2 (Mar./Apr. 1972):75-77.
        A reprint of entry C121.

C138   "Peking's New Petro-Chemical Complex." <u>Peking Review</u> 15, no. 15 (Apr. 14, 1972):4.
        A brief profile of the Peking General Petrochemical works under construction in the southwestern outskirts of China's capital. More than 10,000 workers and soldiers are taking part in the building of the complex. In October 1969 construction began on three refining units with the capacity to process 2.5 million tons of crude oil. By 1972 they had gone on line. When completed, the complex will comprise fifteen refining and petrochemical units capable of turning out a variety of petroleum products, including gasoline, kerosene, diesel oil, lubricating oil, synthetic rubber, polystyrene, and others.

C139   "Peking Builds General Petrochemical Works." <u>Economic Reporter</u> (English supp.) 2 (Apr./June 1972):18-19. illus.
        Describes the construction of the Peking General Petrochemical Works, which started in the winter of 1968, with many of the facilities completed by 1972, including a refinery with an annual processing capacity of 2.5 million tons of crude oil.

C140   "Fuel Industry: Petroleum," by Ladislao La Dany. <u>China News Analysis</u> 885 (June 23, 1972):1-7.
        This profile of China's petroleum industry covers exploration, oil deposits, production, refining, and so forth.

C141   "Mineralogy and Geochemistry of Shelf Sediments of the South China Sea and Taiwan Strait," by Ju-chin Chen. U.N. Economic and Social Commission for Asia and the Pacific, Committee for Coordination of Joint Prospecting for Mineral Resources in Asian Offshore Areas. <u>Technical Bulletin</u> 6 (July 1972): 99-115. illus., tables.
        A tectonic investigation of the offshore hydrocarbon potential in the South China Sea and the Taiwan Strait, which is a part of China's outer continental shelf.

C142   "Sediments of Taiwan Strait and the Southern Part of the Taiwan Basin," by J.T. Chou. U.N. Economic and Social Commission for Asia and the Pacific, Committee for Coordination of Joint Prospecting for Mineral Resources in Asian Offshore Areas. <u>Technical Bulletin</u> 6 (July 1972):75-97. illus., tables.
        Reports on a tectonic study of offshore Taiwan for hydrocarbon potential and prospects.

C143   "A Women's Oil Team--Report from Taching Oilfield," by Ching
       Hung. Peking Review 15, no. 28 (July 14, 1972):19-20. illus.
            A profile of a 100-member female oil production team at
       work in the Tach'ing oil field. Except for 3 men, the team
       is composed of young women who have recently graduated from
       high school at the age of 21. Since the team was formed in
       1970, these highly motivated, unmarried workers have performed
       their jobs as well as men.

C144   "Chinese Refining, Chemical Capacity Climbing." Oil and Gas
       Journal 70 (July 17, 1972):58-59. illus.
            With crude oil production rising 30% annually and reaching
       520,000 b/d in 1971, the expansion of refining capacity is
       also proceeding at a record pace. New refineries are mush-
       rooming in various areas, although their equipment is said to
       be obsolete. A large-scale petrochemical works has been built
       in Peking by 10,000 workers. It has a capacity to handle
       50,000 b/d of crude oil. Ten refining and 3 petrochemical
       units are said to be on line at the end of 1971.

C145   "China: A Bid for U.S. Help in Unlocking Its Oil." Business
       Week, no. 2247 (Sept. 23, 1972):40-41. map.
            To exploit her offshore oil deposits, China needs advanced
       offshore drilling technology. She has purchased a rig from
       Japan and made inquiries through intermediaries about the
       purchase of U.S. offshore equipment. Business Week reasons
       that the U.S. oil industry possesses the most advanced tech-
       nology and is best suited to provide aid if help is requested.

C146   "Taching Oilfield's Fresh Victories." Peking Review 15,
       no. 39 (Sept. 29, 1972):23.
            A news summary of Tach'ing's gains since production
       started in 1961. An annual increase of 35% in crude oil pro-
       duction has been achieved in the past 11 years. A gain of 20%
       or more has been made in the first 8 months of 1972, as com-
       pared with the same period in 1971. Also, all-round improve-
       ments in the quality and cost of petroleum products have been
       attained.

C147   "Openings in China." Petroleum Press Service 39, no. 10
       (Oct. 1972):359-61.
            Comments on the passing of China's self-imposed isolation
       and her new open-door policy toward the West regarding the
       exploitation of her oil deposits.

C148   "Chinese Bottlenecks," by Leo Goodstadt. Far Eastern Economic
       Review 78, no. 41 (Oct. 7, 1972):39-41. illus., map.
            Although China has made major strides in the production of
       petroleum, several negative factors still keep her petroleum
       industry from attaining a more advanced level. These include
       a lag in oil refining capacity, poor quality of crude oil,

inferior oil equipment, a lack of advanced technological expertise, and the generally low quality of petroleum products. Goodstadt concludes that the importation of sophisticated technology from the West is a "must" if China is to surmount these constraints.

C149   "China: Is the Name of the Game Oil?" Forbes 110 (Oct. 15, 1972):29-30. illus.
        In the wake of President Nixon's visit to China, Forbes speculates on the prospects for sales of Chinese oil to the U.S. and for substantial Chinese orders of American oil technology. Although China's current crude oil output is estimated at only 30 million tons, chances are good for major Chinese purchases of U.S. oil equipment in exchange for Chinese oil.

C150   "Tap Potentials, Increase Production," by the Chinese Communist Party Committee of the General Petrochemical Plant, Taching Oil Field." Selections from China Mainland Magazines 739-40 (Oct. 30-Nov. 6, 1972):44-48. Translated from Hung ch'i, no. 10 (Oct. 1, 1972).
        An article from China's top party journal enunciating China's early method of encouraging production through ideological indoctrination and mass enthusiasm and not material incentives.

C151   "China's Crude Output Still Climbing." Oil and Gas Journal 70 (Nov. 6, 1972):49. illus.
        There have been substantial gains in China's output of crude oil during 1971, estimated at 510-520,000 b/d, as compared with the 1972 target of 600-650,000 b/d. There also was a steep drop in China's purchase of oil equipment and a complete halt in her import of petroleum products from the Soviet Union.

C152   "Hercules to Taiwan." Chemical Week 111 (Nov. 22, 1972):27.
        Hercules, an American corporation, may provide technology for the construction in Kaohsiung of a $16 million polypropylene plant, which will have an annual capacity of 50,000 tons. When operational, this Taiwanese plant would also produce low and high density polyethlene.

C153   "Petroleum Industry--China's Economy at a Glance." Peking Review 15, no. 47 (Nov. 24, 1972):17-18. illus.
        Reports on China's oil industry during the Cultural Revolution. Crude oil production in 1970 topped 1969 by 40.9% and a gain of 28.6% was made in 1971 over 1970. Big strides have been made on the expansion and building of new oil refineries, including China's first large-scale integrated petrochemical complex--the Peking General Petrochemical Works. A complete line of petrochemical products has been made

available. Also, major advances have been achieved in oil
exploration.

C154   "China's Oil Industry Develops Rapidly." Economic Reporter
       (English supp.) 4 (Oct./Dec. 1972):20-21. illus.
            This summary of China's achievements during the past year
       reports that the Chinese have succeeded in building a rela-
       tively sophisticated, integrated petroleum industry that in-
       cludes oil prospecting, extraction, transportation, and
       refining.

C155   "Oil Bases in the Desert--Journey to Tsaidam Basin (I)."
       China Reconstructs 22, no. 1 (Jan. 1973):22-27. illus. (some
       color).
            A rare glimpse of the vast Tsaidam basin in Tsinghai
       province and oil exploration operations in the area. The
       basin lies at 2,500-3,000 meters above sea level and covers
       an area of 300,000 square kilometers in northwest Tsinghai
       province. Oil exploration has been conducted in high moun-
       tains and on sand dunes. Some fileds were discovered in 1958
       and a refinery has been built.

C156   "The Petrochemical Industry of Taiwan--Raw Materials," by J.J.
       Dorsey. Industry of Free China 29, no. 1 (Jan. 1973):13-27.
            A primary survey of the petrochemical industry in Taiwan
       with an emphasis on its raw material--natural gas. The avail-
       ability of domestically produced natural gas for industrial
       use has contributed to the growth and development of petro-
       chemicals in Taiwan from 1963 to 1973. From a modest invest-
       ment by Mobil China Allied Corp. in 1963, multinational com-
       panies have since increased their substantial stake in
       Taiwan's petrochemical industry in Taiwan, making it a
       significant factor in the industry of East Asia.

C157   "Fast Boat to China (Polypropylene Resin)." Chemical Week
       112 (Jan. 31, 1973):15.
            Sobin Chemical (Boston) is preparing to make the first
       direct shipment of U.S. industrial chemicals in China since
       1949. President Julian Sobin signed a contract at the Canton
       Trade Fair for the sale of 500 tons of polypropylene and 2,000
       tons of resin, to be shipped directly from the U.S. to China.

C158   "China: Export of Crude Oil." Petroleum Press Service 40,
       no. 2 (Feb. 1973):69.
            Reports on China's initial offer to sell crude oil to
       Japan at the rate of one million tons each year.

C159   "Development of Man-Made Fibre Industry and Petrochemical
        Industry in Taiwan," by H.H. Chen. Industry of Free China
        39, no. 2 (Feb. 25, 1973):18-34. tables.
             Describes and reviews the building of Taiwan's petro-
        chemical industry, which has been upgraded and now includes
        synthetic fiber manufacturing. While using petrochemical
        products as feedstocks, the industry seems to have greater
        growth potential.

C160   "Mainland China's Oil Growth Rate Dips Sharply in 1972." Oil
        and Gas Journal 71 (Feb. 26, 1973):28-29. illus.
             Reports on a slowing of the growth rate of oil production
        in China in 1972. The rate is estimated to be only 5%, a
        sharp decline from a gain of 16% in 1971 over the previous
        year. The Journal estimates China's crude oil production in
        the current year to be on the order of 600,000 b/d.

C161   "China's Petroleum Industrial Front Scores Substantial
        Achievements." Chinese Economic Studies 6, no. 3 (Spring
        1973):8-12. Translated from Jen-min jih-pao (People's Daily)
        (Oct. 6, 1971).
             Summarizes the gains made by China's oil industry during
        the first 8 months of 1971, when there was a 27.2% increase
        in crude oil production and 33.7% growth in refined products.
        Almost all sectors in the oil industry--including exploration,
        drilling, oil field development, and refining--registered
        gains during that period.

C162   "12 Glorious Years of Taching Oilfield." Economic Reporter
        (English supp.) 1 (Jan./Mar. 1973):28-29. illus.
             A historical sketch of the growth and achievements of
        China's largest oil field, Tach'ing, since its discovery and
        development in 1959-1960. The article stresses China's self-
        reliance in exploring and developing its oil base, as well as
        building refining and petrochemical facilities at Tach'ing.

C163   "Mainland China, Russia Oil Search in Infancy," by Louis G.
        Weeks. World Oil 176 (Apr. 1973):94-95. illus.
             Discusses the lagging oil technology of the USSR and
        China, despite some success by the latter in exploring oil-
        bearing basins (especially Sungliao basin) in many areas of
        China. Currently, China's oil and gas prospects have not been
        fully appreciated. China has a very high percentage of sedi-
        ments in many basins, which are made of continental facies.
        Oil equipment used in China is primitive by U.S. standards.

C164   "On the Taching Oilfield," by Wilfred Burchett. Eastern
        Horizon 12, no. 4 (Apr. 1973):9-19. illus.
             With many colorful photos, the Australian journalist de-
        scribes the Tach'ing oil complex in operation and vividly
        portrays the myriads of people at work in the oil fields and
        on the farm.

C165   "China: Development of the Oil Industry." <u>Far East Trade and</u>
       <u>Development</u> (London) 28, no. 4 (Apr./May 1973):156-58.

C166   "First in China Trade (Acrylonitrile)." <u>Chemical Week</u> 112
       (May 16, 1973):23.
            Standard Oil of Ohio has agreed to license the use of its
       process for producing acrylonitrile. This is the first time
       that China has agreed to pay royalties to a U.S. firm.
       Sohio's process has been applied in many countries to the
       manufacture of such synthetic fibers as Orlon and Acrilon.
       The duration of the license and the amount of money involved
       have not been made public.

C167   "New Type Socialist Oilfield." <u>Economic Reporter</u> (English
       supp.) 2 (Apr./June 1973):24-25. illus.

C168   "Rapid Expansion of Petroleum Industry." <u>Peking Review</u> 16,
       no. 39 (Sept. 28, 1973):22.
            Reports on a big gain registered in China's oil industry
       in 1972, with crude oil output reaching 4 times that of 1965.
       Also, oil finds were reported in China's offshore areas. The
       achievements in the third quarter include the commissioning of
       100 new conventional and water-injection wells.

C169   "Two French Firms to Build Chinese Petrochemical Works." <u>Oil</u>
       <u>and Gas Journal</u> 71 (Nov. 5, 1973):34.
            A contract worth $300 million for the construction of a
       large petrochemical complex has been signed between two French
       firms and the Chinese. Technip and Speichim are the prime
       contractors for the production of synthetic fibers and plas-
       tic. The location of the plant has not been disclosed. Con-
       struction will start in spring 1975, and production will be
       developed in line with the steam cracking and catalytic re-
       forming process of the French Petroleum Institute. The project
       calls for 500 French engineers and technicians to work in
       China.

C170   "Tapping Natural Gas in Szechuan." <u>Peking Review</u> 16, no. 47
       (Nov. 23, 1973):23.
            Reports on the rapidly expanding natural gas industry in
       the province of Szechuan. In the past few years, a dozen new
       gas fields were developed and came on line in southwest
       Szechuan and a 1,000-kilometer gas pipeline was laid. The
       production of natural gas in the province in 1973 tripled that
       of 1965.

C171   "China's Fast-Growing Oil Industry," by Hsin Chung. <u>Economic</u>
       <u>Reporter</u> (English supp.) 4 (Oct./Dec. 1973):24-25. illus.
            This brief historical review of China's petroleum industry
       emphasizes gains registered in virtually every area, including
       drilling, production, refining, petroleum products, and petro-
       leum technology.

C172   "Taiwan's Petrochemical Industry in a World Context."
       Economic Review (International Commercial Bank of China,
       Taipei), no. 156 (Nov./Dec. 1973):1-9. tables.
           Reviews Taiwan's fledgling petrochemical industry in the
       context of worldwide supply and demand. It also provides
       statistical data on a wide variety of petrochemical products
       manufactured in Taiwan.

C173   "The Rise of China's Oil Industry." Swiss Review of World
       Affairs 23, no. 9 (Dec. 1973):8-9. illus.
           Summarizes in capsule form the growth and development of
       China's petroleum industry from 1947 to 1973.

C174   "Drilling Pushed in China's Pohai Gulf." Oil and Gas Journal
       71 (Dec. 31, 1973):74. map.
           Describes China's offshore explorations and drilling
       activities in the Pohai Gulf near Tientsin. Three jackup
       rigs--one bought from Japan and two apparently built by the
       Chinese themselves--are reportedly operating in 66 to 100 foot
       deep waters. China is also said to have ordered $20,000,000
       worth of offshore logistics vessels from Denmark and to have
       purchased two offshore rigs from Mitsubishi Heavy Industries
       of Japan.

C175   "China." In Energy and U.S. Foreign Policy, by Joseph Yarger
       and Eleanor Steinberg. Cambridge, Mass.: Ballinger Press,
       1974, pp. 209-27. tables.
           Reviews China's overall energy policy with particular
       attention to production and consumption of fossil fuels such
       as coal, oil, and gas.

C176   "Petrochemicals," by Harned Pettus Hoose. In Doing Business
       with China: American Trade Opportunities in the 1970's, by
       William W. Whitson. New York: Praeger, 1974, pp. 312-23.
           A brief introductory survey of China's petrochemical
       industry, with emphasis on chemical facilities and equipment
       that China needs to modernize her petrochemical industry.

C177   "New Oilfields Opened." Peking Review 17, no. 1 (Jan. 4,
       1974):5.
           Reports from China indicate that many new oil and gas
       fields have been discovered in several provinces, municipali-
       ties, and autonomous regions. Locations of the oil fields
       have not been disclosed, but all the major ones, such as
       Tach'ing, Shengli, and Takang, have scored substantial gains
       in crude oil production.

C178   "China Is Increasing Oil-Export Deals," by Joseph Lelyveld.
<u>New York Times</u> (Jan. 5, 1974):35. illus.
    Because of the world crisis caused by the 1973 Arab oil
embargo, oil prices have recently jumped more than threefold.
China is taking this opportunity to make a new commitment to
export a modest volume of oil. It is not China's current
output, however, but rather the vast potential of her offshore
area that excites the imagination of Western oilmen.

C179   "Chinese Tap Taching Potential," by Wilfred Burchett. <u>Far
Eastern Economic Review</u> 83 (Jan. 14, 1974):45-46. map.
    An eyewitness report about the Tach'ing oil field, the
author's impressions, and his predictions about Tach'ing's
contributions to China's export potential. Burchett describes
the unique Chinese method of oil field development by mobiliz-
ing massive manpower for drilling and maintenance. The poten-
tially vast resources of this area make China an ideal market
for the exporters of oil drilling equipment and petrochemical
facilities.

C180   "China's Oil," by Nicholas Ludlow. <u>U.S.-China Business Review</u>
1, no. 1 (Jan./Feb. 1974):21-27. illus.
    An analysis and evaluation of China's oil strategy. Be-
cause of deteriorating relations with the USSR, China wants to
cooperate with Japan by tapping and selling oil to her. It is
also possible that Sino-U.S. deals on oil could change American policy toward Taiwan. And China's oil exports to Southeast Asian countries can promote economic and political relations. Above all, oil exports can greatly improve China's
balance of payments situation. The author predicts that by
the year 1985: (1) China will possess substantial offshore
expertise and equipment; (2) the U.S. will aid her in offshore
technology; (3) China will own a sizable tanker fleet;
(4) U.S.-China trade will jump because of China's oil exports;
and (5) China will become a world economic power because of
her oil.

C181   "China's Oil Export to Asia." <u>U.S.-China Business Review</u> 1,
no. 1 (Jan./Feb. 1974):22.
    This brief article documents and chronicles China's oil
exports since April 1973. After an initial one-million-ton
crude oil export contract signed with a Japanese trading firm
in 1973, China has concluded subsequent oil deals with
Thailand for the sale of 50,000 tons of diesel fuel and has
boosted by 20% her supply of diesel oil to Hong Kong. Barter
trade involving the exchange of Chinese oil for Philippine
minerals, primarily copper, is said to be progressing well.

C182   "China's Oil Production and Consumption," by Peter Weintraub.
        U.S.-China Business Review 1, no. 1 (Jan./Feb. 1974):29-31.
            A general survey of China's oil reserves, production, and
        consumption in 1973. China's onshore reserves as of 1973 are
        calculated to be 6-10 billion tons; her crude oil output in
        1973 is estimated at 44 million tons; while her refining
        capacity is put at 31 million tons. Offshore and shale oil
        deposits are thought to be around 20 billion tons each as of
        1973.

C183   "Life at an Oil Refinery near Peking," by H.F. Marks. U.S.-
        China Business Review 1, no. 1 (Jan./Feb. 1974):28.
            A brief portrayal of the Peking General Petrochemical
        Works. The construction of the plant started in 1968 with
        10,000 workers, and by the end of 1973, major portions have
        been completed. It has a total of 35 units of refining and
        petrochemical units, of which 11 are on line, including an
        oil refinery capable of treating 2.5 million tons of crude
        oil. The complex employs 21,200 workers, with about 80,000
        people (20,000 families) living in the surrounding area.

C184   "The Oil Question." China Trade Report 1 (Feb. 1974):8.
            In the midst of Arab oil embargo, the Report discusses
        and speculates on how China might be tempted to capitalize on
        this opportunity to export crude oil abroad. Quoting reports
        from the New York Times and the Financial Times, the Report
        maintains that both the internal and external climate appears
        conducive to China's becoming a factor in the oil market in
        Asia. The year 1974 signifies China's first entry into Asian
        market as an oil exporter.

C185   "Practice First: A Chinese News Report." U.S.-China Business
        Review 1 (Jan./Feb. 1974):26.
            An excerpt of translations from Chinese news sources on
        China's oil technology. Reports that from January to May
        1973, technicians and engineers at Tach'ing's Institute of
        Design and Research have assembled much data and completed
        designing of 154 projects, including electricity and water
        supply projects that are advanced by world standards.

C186   "Chou En-lai's Million B/D." Petroleum Economist 2 (Feb.
        1974):49-50. map.
            Analyzes and comments on Premier Chou En-lai's revelation
        that China produced 50 million tons of crude oil in 1973.
        Since Chou En-lai told Edgar Snow that China's petroleum out-
        put stood at 20 million tons in 1970, the 50-million-ton fig-
        ure just 3 years later is both astonishing and confusing. The
        Economist tentatively concludes that the Tach'ing, Shengli,
        and Takang oil fields contributed the lion's share of gains in
        1973.

C187   "Chinese Oil Lures Far East Customers." Oil and Gas Journal
        72 (Feb. 25, 1974):84-86. illus., map.
            Reports on the substantial gains in China's crude oil pro-
        duction during the past few years. A figure of 50 million
        tons was disclosed by Premier Chou En-lai. As a result, China
        is conducting a brisk oil export trade with such Far Eastern
        buyers as Hong Kong, Japan, and Thailand. Accelerated ex-
        ploration is reported in many provinces of China, such as
        Kwangtung, Chekiang, Hupeh, and Kweichow, with possible off-
        shore oil finds. Also reported is a significant boost in
        China's petroleum refining capacity--from 400,000 b/d in 1972
        to an estimated 800,000 b/d in 1974.

C188   "Energy in the People's Republic of China," by Genevieve C.
        Dean. Energy Policy 2, no. 1 (Mar. 1974):33-35.
            Offers an overall review of China's coal, electric power,
        and oil industries. Attempts to evaluate the implications of
        China's political posture and economic policies on the inter-
        national scene in light of her emergence as a petroleum pro-
        ducer. For the foreseeable future, China can be expected to
        be only a modest exporter of energy, particularly petroleum.
        She is also expected to purchase advanced Western technology
        and knowhow to modernize her oil industry and develop her
        growing petroleum resources.

C189   "Inside Look at China's HPI," by Sy Yuan. Hydrocarbon Process
        53 (Apr. 1974):105-8. illus., map.
            Yuan presents his overall impressions and firsthand obser-
        vations of the chemical and petroleum processing industry in
        China during a month-long visit in June-July 1973. He also
        briefly describes the Shanghai and Fushan No. 2 oil refin-
        eries. The author concludes that China is on her way to
        accelerating and expanding petroleum processing, primarily by
        improving the technological level of existing plants and by
        building new ones that are equipped only with advanced
        machinery.

C190   "Peiping's Petroleum Industry: Growth and Future Develop-
        ment," by Chun Chang. Issues and Studies 10, no. 8 (May
        1974):41-56. tables.
            Paints a rather pessimistic picture of China's oil re-
        serves, production, refining, and import of oil technology.
        The author outlines the negative aspects of China's oil pro-
        duction in the future and questions Peking's claim of attain-
        ing an oil output of 54 million tons in 1973, insisting that
        she could have produced no more than 25 million tons.

C191   "Newly Built Takang Oilfield--Paean to the Great Cultural
        Revolution." Peking Review 17, no. 21 (May 24, 1974):15-17.
        illus.
            A portrayal of the Takang oil field from its discovery in
        1964 to its subsequent development into a major oil base by
        1974. Located in northeast China, the Takang field was ex-
        ploited during the Cultural Revolution. Since 1967, crude oil
        output has been increasing at 60.9% per annum.

C192   "Fruit of Great Cultural Revolution: Taching Is Five Times
        Its Former Self." Peking Review 17, no. 23 (June 7, 1974):
        16-18. illus.
            Compares the Tach'ing oil field in 1965 with that of 1973
        in terms of production and construction. Tach'ing has grown
        fivefold, with the original investment providing a return of
        1,100% in accumulated surplus funds. Tach'ing has also ex-
        panded its industrial complex and has branched out into the
        manufacture of synthetic ammonia, ammonium nitrate, and
        acrylonitrile.

C193   "New Type Oilfield." Peking Review 17, no. 24 (June 14, 1974):
        23.
            A story of how oil field workers and their families in
        Yümen, Kansu province, have succeeded in developing agricul-
        tural settlements alongside the oil industry in a barren
        wasteland of northwest China. They learned from the examples
        of Tach'ing and reclaimed 133 hectares of useless land, estab-
        lishing 28 farms and producing grain in that remote region.

C194   "The Petrochemical Industry in Taiwan--Overall Planning."
        Industry of Free China 41, no. 6 (June 25, 1974):13-26.
            An excerpt from The Outlook for the Petrochemical Industry
        in Taiwan, prepared by Arthur D. Little Co. This is a profile
        of petrochemical section of Taiwan's manufacturing economy in
        the overall context of worldwide investment trends in petro-
        leum and petrochemicals. The article describes the charac-
        teristics of Taiwan's petrochemical industry, including its
        production data and competitive position in the world.

C195   "Taiwan's First Styrene Plant Ordered." Petroleum Inter-
        national 7 (July 1974):36.
            Badger of Great Britain has signed a contract with Delta
        Petroleum Corp. of Taipei for the construction of an ethyl-
        benzene/styrene plant to be built at Ta She petrochemical com-
        plex near Kaohsiung in Taiwan. This is the first styrene
        plant to be built in Taiwan.

C196 "Gas for Industry and Home Use." <u>Peking Review</u> 17, no. 27
(July 5, 1974):31.
The greater availability of natural gas has resulted in an
expansion of gas supplies for residential users. In contrast
with 1965, when only 16 cities had access to natural gas,
residents of 31 Chinese cities are now supplied with natural
gas for cooking. The noncommercial use of gas has more than
doubled in the past 10 years, while the number of consumers
has tripled. Gas is currently available in most major Chinese
cities.

C197 "More Crude Oil." <u>Peking Review</u> 17, no. 34 (Aug. 23, 1974):
23.
Crude oil production in the first half of 1974 has shown a
gain of 21.3% over the same period in the previous year, while
refined petroleum products have increased by 24.7%. All key
oil fields have shown increases ranging from 24.7 for Tach'ing,
12.2% for Shengli, and 22.5% for Takang. Also, several new
fields have been developed and placed in production.

C198 "China's Petroleum Industry Is Booming and Overfulfills Its
1973 Production Quotas." <u>Chinese Economic Studies</u> 8 (Fall
1974):58-60.
This article, a translation of a story that first appeared
in <u>Jen-min jih-pao</u> (People's Daily), provides an account of
the operation and performance of China's oil industry in 1973.
Production of crude oil and natural gas went up substantially,
and the Chinese have made progress in other areas, such as
exploration, drilling, and refining.

C199 "China Opens New Oilfield." <u>Economic Reporter</u> (English
supp.) 3 (July/Sept. 1974):30-31. illus.
A brief account of the development of the Takang oil field
between 1967 and 1973. Crude output at China's third largest
field has been increasing at a compounded rate of 60.9% since
1967. A contingent of 10,000 workers from Tach'ing is primarily responsible for the opening of this new field.

C200 "A Major Exporter?" <u>Petroleum Economist</u> 9 (Sept. 1974):335.
Comments on the statement issued by Mr. Hasegawa, chairman
of the Japan-China Oil Import Council, that China's crude oil
output will rise to 400 million tons in a near future. The
<u>Economist</u> regards the statement as too sanguine to be realis-
tic, but observes that with the annual gain in crude oil output at more than 20% in the past few years, it would not be
unreasonable for Japan to obtain Chinese oil on the order of
100 million tons a year.

C201   "Soviet Union Refuses to Concede China Has Huge Offshore
        Reserves," by Donald R. Bakke. Offshore 34, no. 10 (Sept.
        1974):64-65.
            The USSR denies a report that China has vast oil deposits
        off its northeast coast. The Soviets are particularly upset
        over the story that China's offshore oil reserves along her
        northern coast can match or surpass those of Saudi Arabia,
        since competition from such huge reserves could undermine the
        Soviet strategy of influencing oil-hungry Asian countries by
        supplying them crude oil.

C202   "Conoco Hits Gas and Condensate off Taiwan." Oil and Gas
        Journal 72, no. 35 (Sept. 2, 1974):28. map.
            Continental Oil Co. announced a major gas and condensate
        discovery off the coast of southern Taiwan. The flow was
        estimated at 25,000 Mcf. of gas and 250 b/d of oil-like
        condensate.

C203   "Chinese Market for Oil Exploration Gear Is Tapped Cautiously
        by U.S. Companies," by D'Arcy O'Connor. Wall Street Journal
        48 (Sept. 6, 1974):26.
            Discusses the views of U.S. oil equipment manufacturers
        about their prospects in the Chinese market. Although annual
        sales are currently at a modest amount of $10 million, China
        could greatly increase her orders for advanced American oil
        equipment in the future, since she recognizes the enormous
        potential for foreign-exchange earnings through the develop-
        ment of her oil resources using foreign technology.

C204   "China's Petroleum Industry--An Enigma," by Horton R. Connell.
        Bulletin of the American Association of Petroleum Geologists
        58, no. 10 (Oct. 1974):2157-72. illus., maps.
            The most comprehensive and detailed analysis and survey of
        China's petroleum industry to appear in the Bulletin since
        Meyerhoff's article in 1970 (entry C99). Connell attempts to
        follow up the development of the Chinese petroleum industry
        since 1968. He examines Chinese government policies relating
        to the petroleum industry and updates the rate of China's
        crude oil production from 1968 to 1973. The author also pro-
        vides estimates of crude oil production from individual oil
        fields, attributing overall gains primarily to output from
        Tach'ing, Shengli, Takang, and the Karamai-Urho fields of
        Sinkiang province.

C205   "How China Developed Her Oil Industry," by Chun Chang. China
        Reconstructs 23, no. 10 (Oct. 1974):2-7. illus.
            A chronological and historical profile of the development
        of China's petroleum industry from its inception to 1974,
        emphasizing its postwar growth. In describing the state of
        backwardness before 1949, the author credits China's recent
        successes to the correct leadership of the Communist Party and

to the spirit of independence, self-determination, and self-reliance.

C206   "New Oilfields Go into Production." China Pictorial 10
        (Oct. 1974):2-5. illus. (some color).
             Illustrates with colorful photos the development of
        China's new oil bases in the past ten years. Starting with
        Tach'ing oil field in 1960, China succeeded in completing
        other major fields, such as Shengli in 1964 and Takang in 1967.

C207   "A Visit to Takang Oilfield." China Reconstructs 23, no. 10
        (Oct. 1974):8-14. illus.
             A firsthand report on visits to the Takang oil field. De-
        scribes the growth and development of Takang since its dis-
        covery in 1964 and includes heroic stories of the drilling
        teams surmounting difficulties and striking rich wells from
        erratic geological strata.

C208   "Outlook Bright for Oil, Coal--China '74 Focus," by Christopher
        Lewis. Far Eastern Economic Review 86, no. 39 (Oct. 4, 1974):
        21-22. illus., graphs.
             A look back on China's energy situation and development in
        early 1974 and late 1973, with the emphasis on petroleum.
        The author assesses China's estimated oil reserve, analyzes
        her current oil production and prospective output, and ap-
        praises her initial exports to neighboring countries, includ-
        ing Japan. He concludes that the outlook for China's oil
        industry is highly promising and is especially enthusiastic
        about China's prospect of developing her vast offshore oil
        deposits and the concomitant purchase of offshore technology
        from Western nations.

C209   "Mainland China Claims First-Half Oil Production Up 21.3%."
        Oil and Gas Journal 72 (Oct. 7, 1974):52.
             Reports on China's claim of significant gains made in
        crude oil output and expansion of refining capacity, with
        crude up 21.3% and 18.2%, respectively, from January to June
        1974. Crude oil production at Tach'ing, Shengli, and Takang
        in the period went up by sizable margins.
        Soviets show skepticism about the quantity of China's offshore
        oil production.

C210   "Shengli Oilfield Thrives--New Achievement in Socialist Con-
        struction." Peking Review 17, no. 41 (Oct. 11, 1974):16-17.
        illus.
             A brief history of this major oil base on the Pohai Gulf
        from its inception in 1964 to its development into the second
        largest and one of the most productive oil fields in China.
        No details on oil reserves or production figures are given,
        but most wells are reported to be high-yielding. The location
        of this oil field was later identified as north of Shantung
        province and south of the Pohai Gulf.

C211   "Matter of Semantics (Gulf of Pohai)," by Frank J. Gardner.
        Oil and Gas Journal 72 (Oct. 21, 1974):91.
            Discusses and speculates on the geographical definition
        of the Takang oil field. Since no foreigner has ever visited
        Takang, it is shrouded in secrecy. The author tries to probe
        whether it is an onshore or offshore oil base. The field is
        said to be covered with shallow water only a few feet deep,
        but China has never claimed Takang as an offshore find. It
        is obviously on the border of the Pohai Gulf.

C212   "Explorations: Another Player in the Oil Game," by W. Glenn.
        Far Eastern Economic Review 86 (Oct. 25, 1974):59.
            Comments on oil and gas discovered offshore to the south-
        west of Kaohsiung in Taiwan.

C213   "An Appraisal of the Changing Fuel Structure on the China
        Mainland," by Yu-chiao Hung. Issues and Studies 10, no. 14
        (Nov. 1974):58-76. tables.
            Hung examines the various estimates of China's crude oil
        production and relates her oil supply to total energy consump-
        tion. He tentatively concludes that in spite of China's
        claims to have boosted production and developed such large oil
        fields as Tach'ing, Shengli, and Takang, petroleum and natural
        gas provide only 20% of total energy used. The balance is
        provided by coal. Hung therefore disputes China's claim to
        have made substantial gains in crude oil output and to have
        surplus oil available for export.

C214   "China's Forward Leap." Petroleum Economist 41, no. 11
        (Nov. 1974):408-10. map.
            Comments on and assesses China's great leap forward in the
        production of crude oil, oil processing, and oil exports. The
        confirmation of China's crude oil production at a 50-million-
        ton level in 1973 is viewed as very significant, heralding
        China as an emerging oil power. With oil output likely to
        reach 100 million tons by 1976 and 400 million tons by 1984,
        China has become an important factor in the energy world.

C215   "Giant Gas Tank." Peking Review 17, no. 47 (Nov. 22, 1974):
        23. illus.
            Reports that China's largest gas storage tank, with a
        capacity of 150,000 cubic meters, has been built in Shanghai.
        This 12-story-high structure is owned and operated by the
        Shanghai Gas Co. Two smaller tanks with respective capacities
        of 5,000 and 20,000 cubic meters were completed some years
        ago.

C216   "The Energy Situation in Taiwan, Republic of China--Past, Present, and Future," by K.S. Chang. Industry of Free China 5 (Nov. 25, 1974):2-16. Economic Review 162 (Nov./Dec. 1974): 1-13.
       Summarizes the development of Taiwan's energy resources from 1954 to 1973. The author briefly discusses the 10% annual growth in energy demand in the past 10 years, outlines some measures to tackle a sharp rise in imported oil prices, and examines alternatives to alleviate Taiwan's heavy dependence on imported oil. Gains in the production of natural gas at home and problems of offshore drilling for oil are examined. Finally, Chang foresees a more efficient use of energy through the development of the petrochemical industry in Taiwan.

C217   "China's Taching Oilfield Expanded at Top Speed," by Hung Yu. Economic Reporter (English supp.) 4 (Oct./Dec. 1974):30.
       An account of the major gains achieved by workers at the Taching oil field. Crude oil output increased 26% during the first 4 months of 1974, despite the transfer of over 10,000 veteran workers from Tach'ing to new assignments at the Takang oil field.

C218   "Petroleum Industry in China," by Tatsu Kambara. China Quarterly 60 (Oct./Dec. 1974):699-719. maps, graphs, tables, charts.
       One of the more illuminating assessments of China's petroleum industry (1949-1973), by a Japanese oil specialist. Frequently quoted and referred to in many subsequent books and articles. Offers informative studies of China's oil industry from its historical background--primitive conditions in the pre-1949 era through the years of foreign aid (chiefly Soviet), and the period of self-reliance since 1960. In assessing the current situation, the author gives eleven various estimates of crude oil production (including shale oil) from 1960-1973. He gives sketches of individual oil fields and describes the development of refining and the structure of demand at home. Future prospects are termed bright and promising.

C219   "Shengli, China's Second Large Oilfield." Economic Reporter (Eng. supp.) 4 (Oct./Dec. 1974):31. illus.
       Primarily photo illustrations of the Shengli oil field, with oil wells dotting the farmland and scenes of the Shengli Petrochemical Complex.

C220   "Big Hopes for China's Energy Resources." China Trade Report 12 (Dec. 1974):8-9. illus.
       Surveys China's energy resources, placing special emphasis on oil at the end of 1974. Maintains that the sharp international price rise during 1974 has created a golden opportunity for Chinese oil exports. As a result, China is expected to ship 6-7 million tons of crude oil overseas in 1974 (up from

2-3 million tons in 1973). The <u>Report</u> estimates China's crude
oil output in 1974 to be 65 million tons (up from 53 million
tons in 1973) and predicts that it may reach the 200-million-
ton mark in 1980. With such rosy prospects, China has begun
to shop around in world markets for oil technology and equipment that will enable her to accelerate the development of
her vast resources.

C221   "Oil Provides China's Solution," by Leo Goodstadt. <u>Far East-</u>
<u>ern Economic Review</u> 86, no. 49 (Dec. 13, 1974):53. illus.
        Reports that China's balance of payments is expected to
show a deficit of $700 million in 1974. The logical solution
to this problem is the export of crude oil, with China's oil
output estimated at 65 million tons in 1974. Japan has already bought $243 million worth of oil in the first 9 months
of 1974, and prospects for more exports seem promising, as
word circulates that new oil has been found in southern China.

C222   "China and Offshore Oil: The Tiao-yu Tai Dispute," by Victor
H. Li. In <u>China's Changing Role in the World Economy</u>, edited
by Bryant G. Garth. New York: Praeger, 1975, pp. 143-62.
maps.
        Originally appeared in the Spring 1975 issue of <u>Stanford</u>
<u>Journal of International Studies</u> (entry C239). The author
attempts to judge the territorial disputes over oil-rich
islands, involving China, Japan, and Taiwan, on the basis of
international legal precedents.

C223   "China's Energy Policies," by Marcel Toussaint. In <u>Energy,</u>
<u>Inflation, and International Economic Relations</u>, by Curt
Gasteyger et al. New York: Praeger, 1975, pp. 51-57.
        Discusses China's domestic and international energy poli-
cies and the outlook for the development of her petroleum and
other energy resources.

C224   "China's Petroleum Potential," by A.A. Meyerhoff. <u>World</u>
<u>Petroleum Report</u> 21 (1975):18-21. illus., map.
        A reassessment of China's oil and gas reserves, updating
and upgrading the author's previous article "Developments in
China, 1949-1968" (entry C99). Puts China's total onshore and
offshore oil reserves at 70,000 million barrels and natural
gas deposits as 25 trillion cubic feet.

C225   "The Chinese Petroleum Industry: Growth and Prospects," by
Bobby A. Williams. In <u>China: A Reassessment of the Economy</u>
by Joint Economic Committee, U.S. Congress. Washington, D.C.:
Government Printing Office, 1975, pp. 225-63. maps, tables.
        One of the most quoted publications on the petroleum in-
dustry in China. It extensively analyzes many facets of the
industry and offers a chronological survey covering the pre-
1949 years to 1949-1952; the first five-year plan, 1953-1957;

the second five-year plan, 1958-1962; and 1963 to the present (1974). It reports on China's current exploration efforts, petroleum imports and exports, consumption patterns, oil refining, and future trends of output and export potentials. The appendix provides documentation about various oil producing regions, with output estimates for individual fields. It also offers a breakdown of petroleum consumption in China. The paper was originally submitted to the Joint Economic Committee on July 8, 1975.

C226   "People's China and the World Energy Crisis: The Chinese Attitude toward Global Resource Distribution," by Kim Woodard. In China's Changing Role in the World Economy, edited by Bryant G. Garth. New York: Praeger, 1975, pp. 114-42. table.
      Originally appeared in the Spring 1975 issue of Stanford Journal of International Studies. A theoretical analysis of China's domestic and international energy policies, including her prospects for crude oil exports.

C227   "Building Oil Industry through Self-Reliance--A Visit to the New Oil Pipeline as well as Taching and Takang Oilfields (I)," by Jung Hsiang and Hai Yu. Peking Review 18, no. 1 (Jan. 3, 1975):14-16, 29. illus.
      The first installment of a two-part eyewitness report on the building of a 1,152-kilometer pipeline between the Tach'ing oil field and Ch'inhuangtao oil terminal. The authors vividly describe their visit to the bustling oil port of Ch'inhuangtao, followed by a tour of the Tach'ing oil field and an interview with members of an oil drilling team. In addition, this report describes how a "mass battle" was waged by the workers to overcome difficulties in laying the pipeline, which cut across 260 big and small rivers, 40 railway lines, and 200 highways.

C228   "Building Oil Industry through Self-Reliance--A Visit to the New Oil Pipeline as well as Taching and Takang Oilfields (II)," by Jung Hsiang and Hai Yu. Peking Review 18, no. 2 (Jan. 10, 1975):13-16. illus.
      In this second part, the correspondents describe the building of a pipe-making plant in Ssup'ing, the modernization of an oil refinery at Fushun, and the construction of China's third largest oil field at Takang. They emphasize again and again the spirit of self-reliance that drives Chinese workers to build from scratch large oil fields, modern petroleum refineries, and 200,000-ton steel pipes for the construction of a 1,152-kilometer pipeline across several provinces in northern China.

C229   "Ideology and Oil." China Trade Report 13 (Feb. 1975):2.
        Assesses China's efforts to build her own offshore drill-
ing rigs, as well as the prospects for Western sales of oil
technology and equipment to China. In spite of China's success in building of her own jackup offshore rigs, prospects
for selling high-priced oil equipment to China are still
bright, since it is cheaper and more economical to import
foreign facilities than to start building from scratch.

C230   "China Completes Pipeline to Step Up Oil Exports," by
Theodore Shabad. New York Times (Feb. 10, 1975):39.
        China has built her first large-diameter pipeline from the
Tach'ing oil field in northeast China to a tanker port on the
Yellow Sea. The 715-mile-long pipeline can accelerate the
flow of oil from northern Manchuria to ports in Japan. In
1974, China's export of oil to Japan was estimated at 4 million tons, while her total output was about 60 million tons.
Since 1973 China has been exporting crude oil to Japan in
large quantities.

C231   "China's First Long Oil Pipeline Completed." Economic
Reporter (English supp.) 3 (Jan./Mar. 1975):33.
        A brief description of China's first oil trunkline, run-
ning from the Tach'ing oil field to the port of Ch'inhuangtao.
This 1,152-kilometer-long pipeline has 19 pumping stations, at
70 kilometer intervals. Following the completion of the pipeline in 1973, a parallel line was built in October 1974.

C232   "How China's First Long Petroleum Pipeline Was Built."
Economic Reporter (English supp.) 1 (Jan./Mar. 1975):32-33.
illus.
        Outlines the construction of a 1,152-kilometer oil trunk-
line from the Tach'ing oil field to a terminal at the port of
Ch'inhuangtao. Construction commenced in September 1973 and
was completed in less than 2 years.

C233   "China's Petroleum Industry Continues Fast Advance in 1974."
Economic Reporter (English supp.) 1 (Jan./Mar. 1975):30-31.
illus.
        Summarizes the achievements of China's petroleum industry
in 1974. Reports a gain of 20% in crude oil output over 1973,
as well as substantial increases in the production of natural
gas, pipeline building, prospecting, oil field construction,
and oil refining.

C234   "New Achievements in China's Oil Industry," by Ching-yuan Hua.
China's Foreign Trade (Peking) 1 (Jan./Mar. 1975):5-8. illus.
        Discusses the overall performance of China's petroleum
industry, 1963-1974, including the substantial gains in exploration, drilling, production, and refining. Also reviews
the growing market for China's petroleum products.

C235   "Taching's Red Banner Flies Even Higher," by Wen Chung.
        China's Foreign Trade (Peking) 1 (Jan./Mar. 1975):8-12.
            Describes the development and achievements of the Tach'ing
        oil field under the spirit of self-reliance and independence.

C236   "China's First 'Underground Artery.'" China Pictorial 3
        (Mar. 1975):26-27. illus. (some color).
            A colorfully illustrated story depicting the completion of
        a 1,152-kilometer pipeline connecting the Tach'ing oil field
        to the port of Ch'inhuangtao, where Tach'ing crude could be
        shipped by tanker to other parts of China. Pipeline construc-
        tion began in 1970 and was completed in 1975.

C237   "Communist China's Oil Exports: A Critical Evaluation," by
        Vaclav Smil. Issues and Studies 11, no. 3 (Mar. 1975):71-78.
            Critically analyzes China's oil export potential within a
        global context. The author concludes that under current con-
        ditions, China will not be able to produce more than 400 mil-
        lion tons of crude oil by 1990. By then, moreover, rising
        demands from industry and agriculture will severely limit the
        amounts of oil available for export.

C238   "Larger Exports Envisaged." Petroleum Economist 3 (Mar.
        1975):96-97.
            Chinese crude oil exports to Japan in 1975 are expected to
        reach 8 million tons, despite the fact that actual shipments
        of oil in 1974 fell 900,000 tons short of scheduled deliveries.
        China seems determined to make her oil price and quality more
        competitive with other crude oil exporters such as Indonesia.

C239   "China and Off-Shore Oil: The Tiao-yu Tai Dispute," by
        Victor H. Li. Stanford Journal of International Studies 10
        (Spring 1975):143-62. Also in China's Changing Role in the
        World Economy, edited by Bryant G. Garth. New York: Praeger,
        1975, pp. 143-62.
            Li examines and analyzes the controversial claims to the
        potentially oil-rich area surrounding Tiao-yu Tai (Senkaku
        Islands), islands off China's outer continental shelf between
        China and Japan.

C240   "People's China and the World Energy Crisis: The Chinese
        Attitude toward Global Resource Distribution," by Kim Woodard.
        Stanford Journal of International Studies 10 (Spring 1975):
        114-42. table.
            Uses China's societal and structural variables to analyze
        her overall domestic and international energy policies in the
        context of China's participation in the world energy trade,
        including prospects of crude oil exports.

C241   "Petroleum Resources: How Much Oil and Where?" by John D.
        Moody and Robert E. Geiger. Technology Review 5 (Mar./Apr.
        1975):38-46.
            According to Mobil Oil's own research, based on the cur-
        rent knowledge of technology, oil and gas account for only 6%
        of potential available energy resources in the foreseeable
        future. Mobil estimates the world's recoverable oil resources
        at 2 trillion barrels. China, the Middle East, and the Soviet
        Union account for the bulk of potential oil deposits.

C242   "China's First Floating Drilling Vessel for Sea Exploration."
        China Pictorial 4 (Apr. 1975):4-5. illus. (some color).
            A colorful illustration of China's first homemade offshore
        drilling rig, which was built jointly by the Shanghai Hutung
        Shipyard and the Shanghai Mine Drilling Equipment Plant.
        Named K'ant'an No. 1, the rig has already succeeded in drill-
        ing a well in deep waters of the Yellow Sea.

C243   "First Geological Prospecting Ship--New Victories on the Oil
        Front." China Reconstructs 24, no. 4 (Apr. 1975)38-39. illus.
            K'ant'an (Prospector) No. 1, the first floating drill ship
        for marine geological prospecting completely designed and
        built by the Chinese, has completed its first round of trial
        operations for deep water oil prospecting. The hull was con-
        structed by joining two cargo ships together.

C244   "New Victories on the Oil Front." China Reconstructs 24,
        no. 4 (Apr. 1975):36-39. illus.
            (See entries C243, C245, and C247.)

C245   "Pipeline From the Taching Oil Field--New Victories on the Oil
        Front (I)." China Reconstructs 24, no. 4 (Apr. 1975):36-37.
        illus.
            A pipeline, 1,152 kilometers long, extending from the
        Tach'ing oil field to Ch'inhuangtao on the Pohai Gulf has been
        completed. Designed and built in two years by Chinese, the
        pipeline runs through four provinces, Heilungkiang, Kirin,
        Liaoning, and Hopeh, after being commissioned in October 1973.

C246   "A Study of the Energy Sources on the China Mainland," by
        Chung-mon Kang. Issues and Studies 11, no. 4 (Apr. 1975):
        60-74. maps, tables.
            A general survey of China's three major energy industries
        --coal, petroleum, and hydroelectric power. In the section on
        petroleum, the author examines China's effort on oil explora-
        tion, estimates production of crude oil (including shale oil
        output), and discusses prospects for exports.

C247   "Szechuan Gas Fields--New Victories on the Oil Front (II),"
        China Reconstructs 24, no. 4 (Apr. 1975):37-38. illus.
            A news report on gains in natural gas output in Szechuan
        province, up 3.3 times over 1965. Currently, natural gas
        accounts for 2/3 of the fuel for metallurgical plants, 70%
        for nitrogen fertilizer plants, and 83.9% for refining rock
        salt in Szechuan.

C248   "Two New Oil Fields and a New City," by Rewi Alley. Eastern
        Horizon 14, no. 4 (Apr. 1975):9-21.
            Details the impressions of Alley, an American expatriate
        and a longtime resident of China, during his visit to the
        Takang oil field in Tientsin and the Shengli oil field near
        the mouth of the Yellow River in Shantung province.

C249   "Pohai No. 1 Offshore Drilling Rig," by Hsi-shu Hsiao and
        Heng-i Tseng. Selections from the People's Republic of China
        Magazines 816 (Apr. 7, 1975):39-43. Translated from K'o-hsüeh
        shih yen (Scientific Experiment) (Jan. 1975). illus.
            Describes the construction of China's first jackup drill-
        ing rig. This vessel was launched in 1973 and has success-
        fully drilled offshore wells during her trial operations.

C250   "Scientific Experiment Promotes Development of Petroleum Pro-
        duction." Selections from the People's Republic of China
        Magazines 816 (Apr. 7, 1975):30-38. Translated from K'o-hsüeh
        shih yen (Scientific Experiment), (Jan. 1975). illus.
            Describes scientific innovations carried out by Chinese
        technicians to help boost oil production.

C251   "Oil Up, Steel Down?" China Trade Report 13 (May 1975):9.
        illus.
            Speculates on the possibility of a disruption in oil and
        steel production by the February earthquake in the industrial
        province of Liaoning. The quake was severe enough that the
        Tach'ing oil field's pipeline might have been damaged by the
        tremors. The foreign estimate of China's crude oil production
        in 1974 is 60 million tons.

C252   "A Vigorous Force on the Petroleum Front." China Pictorial 5
        (May 1975):18-21. illus.
            Describes the building of Tientsin No. 1 Petrochemical
        Works in progress. The petrochemical facilities being con-
        structed on the outskirts of Tientsin were entirely designed
        and manufactured by Chinese during the Cultural Revolution.

C253   "Szechuan's Gas Fields Making Headway." Peking Review 18,
        no. 19 (May 9, 1975):30. illus.
            An account of the phenomenal development of natural gas in
        Szechuan province, which has made tremendous strides in pro-
        duction since 1949. A single day's output today equals that

of two years in the pre-1949 era. With such an ample supply,
natural gas has become a major energy source for key industries in the province, providing fuel for the metallurgical
industry, raw materials for making nitrogen fertilizer, and
electric power for machine building and salt production.

C254   "Resolutely Follow the Taching Road." Selections from the
People's Republic of China Magazines 823-24 (May 27-June 2,
1975):39-43. Translated from Hung ch'i (May 1, 1975).
      Enunciates Tach'ing's method of success through hard work
and self-reliance, and without material incentive to promote
production.

C255   "The Quantity of Oil and the Quality of Life," by David Crook.
Eastern Horizon 14, no. 6 (June 1975):7-14.
      Depicts a colorful pastoral life in this sprawling oil
city of Tach'ing, which composes not only a petrochemical
complex, but also a thriving rural establishment.

C256   "China's Looking for Oil Too!" Economist 155 (June 28, 1975):
92. map.
      Evaluates the development of the oil industry in China and
ponders the prospects of the China market for British concerns. Considered are China's onshore and offshore oil reserves, her current production and refining capacity, and her
export volumes. To the British, the opportunities for selling
offshore drilling equipment to China seem especially bright.

C257   "China's in the Market for North Sea Know-How," by J. Woward.
Industrial Management (London) (July 1975):34-36.
      Examines the prospects of exporting offshore technology
and equipment to China.

C258   "Oil Guesswork." China Trade Report 13 (July 1975):8.
      Discusses a wide disparity of estimates from specialists
on the output and export potentials of China's oil. Among
those quoted by the Report are Nicolas Ludlow of U.S.-China
Business Review, Hideo Ōno of Asia Quarterly, Prof. Ping-ti Ho
of the University of Chicago in the 1970s, and Masanobu Otsuka
of the Nomura Research Institute in Chūgoku keizai kenkyū
geppō. The Report notes that since these forecasts have not
been backed by hard facts, they can only be considered guesswork.

C259   "Look beyond the North Sea and Sell Hardware to China."
Engineer 241 (July 3, 1975):38-39. illus.
      Discusses the prospects for sales of oil technology and
offshore drilling equipment to China in light of her accelerated effort to develop undersea oil deposits in her outer continental shelf.

C260   "Liquid Assets: China's City of Taching Abounds with Ducks,
        Hogs--and also Oil," by William E. Giles. Wall Street Journal
        (July 7, 1975):1+.
            Impressions of a first visit to Tach'ing. The author is
        rather surprised by the rural outlook of the oil city, with
        its sprawling agricultural settlements around the petrochemi-
        cal complex. He observes that all able-bodied housewives also
        work in the fields.

C261   "China's Oil and Gas Reserves and Resources: A Review."
        Petroleum Times 79, no. 2008 (July 11, 1975):19-35. illus.,
        maps.
            An extensive appraisal of China's petroleum resources,
        both onshore and offshore. Also included are estimates of
        crude oil production capability and refining capacity.

C262   "Chinhuangtao-Peking Oil Pipeline Completed." Peking Review
        18, no. 29 (July 18, 1975):6-7. illus.
            A 1,505-kilometer trunk oil pipeline connecting the
        Tach'ing oil field with the port of Ch'inhuangtao (through
        Peking) was put on line on June 23, 1975.

C263   "China's Energy Resources and Prospects," by Yuan-li Wu.
        Current History 69, no. 407 (July/Aug. 1975):25-27, 53-54.
            An article almost totally devoted to discussing the growth
        of China's oil production since 1965 and its export potentials.
        The author first analyzes China's political motives for ex-
        porting oil to Japan, then examines the historical development
        of oil production in China and export prospects in light of
        the country's overall energy supply and consumption. To
        compute her exportable oil surplus in the future, Wu applies
        the combined growth rates of China's GNP and industrial pro-
        duction to the elasticity of energy demand in the 1970s. He
        finds the amount of China's surplus oil for export not very
        significant.

C264   "China to Initiate Intensive Offshore Oil Exploration--Ocean
        Industry Digest," by J.W. Rasmussen. Ocean Industry 10, no. 8
        (Aug. 1975):141-43. maps, tables.
            An assessment of China's potential to become a major oil
        and gas producer for both domestic supply and export in the
        coming decade, with some emphasis on her offshore oil develop-
        ment.

C265   "Chinese Oil-Industry Image Changing," by Paul H. Fan. Oil
        and Gas Journal 73, no. 32 (Aug. 11, 1975):110-12. map.
            Estimates of the reserves, production, and potential of
        major oil fields in China by an outstanding geologist at the
        University of Houston.

C266　"Mainland China Gearing Up to Boost Oil Exports," by
　　　J. Cranfield. <u>Oil and Gas Journal</u> 73 (Aug. 11, 1975):21-24.
　　　illus., maps.
　　　　　The Sino-British Trade Council says that Peking is aiming
　　　at crude oil production of 3 million b/d by 1980 from both
　　　offshore and onshore wells. It predicts that China's crude
　　　oil exports could reach 400,000 b/d by 1980, since exports in
　　　1974 are already 120,000 b/d, up from 52-60,000 in 1973. The
　　　Council estimates that the crude oil price is currently down
　　　to $12.10 in 1975, from its 1973 high of $14.80 a barrel. In
　　　1974, China exported 1 million tons to North Korea, 800,000
　　　tons to the Philippines, 500,000 tons to North Vietnam,
　　　200,000 tons to Thailand, and 8 million tons to Japan. There
　　　is also major expansion of pipelines and new tanker berths in
　　　northern China.

C267　"China Building a Sizable Oil Industry." <u>Chemical and Engi-</u>
　　　<u>neering News</u> 53 (Aug. 18, 1975):13.
　　　　　Reports and comments on China's petroleum and petroleum-
　　　related fertilizer industry. Quoting a recent report released
　　　by the Joint Economic Committee of the U.S. Congress (entry
　　　C225), the article says that with a more abundant domestic
　　　supply of oil, China began buying ammonia/urea plants, with a
　　　total capacity of 3.78 million tons, using domestic petroleum
　　　as feedstocks. These fertilizer plants are expected to become
　　　operational by 1976. Meanwhile, the increase in crude oil
　　　production enables China to export surplus oil to other coun-
　　　tries, including a projected 8 million tons to Japan in 1975.

C268　"China's First Floating Vessel Completes Maiden Voyage for Sea
　　　Expedition." <u>Economic Reporter</u> (English supp.) 3 (July/Sept.
　　　1975):34-35. illus.
　　　　　China's first offshore drilling rig, <u>K'ant'an</u> (Prospector)
　　　<u>No. 1</u>, has completed trial operations in the southern Yellow
　　　Sea after spudding a well. Built by Shanghai's Hutung Ship-
　　　yard, this floating drilling vessel is made of two Chinese-
　　　made cargo ships "pieced together and reshaped."

C269　"New Pipeline Relays Oil from Taching to Peking." <u>Economic</u>
　　　<u>Reporter</u> (English supp.) 3 (July/Sept. 1975):29. illus.
　　　　　Describes the completion of a new 355-kilometer pipeline
　　　running from the port of Ch'inhuangtao to Peking. The build-
　　　ing of this trunkline will enable China to ship crude oil from
　　　the Tach'ing oilfield to the Tungfanhung Oil Refinery at the
　　　Peking General Petrochemical Works. The pipeline traverses
　　　thirteen counties and cities in Hopeh province.

C270   "Shanghai Builds General Petrochemical Works--New Construc-
        tions." Economic Reporter (English supp.) 3 (July/Sept.
        1975):27.
            Describes construction work in progress at the Shanghai
        General Petrochemical Works in Chinshan county near the city
        of Shanghai. Ten plants for making synthetic fibers have
        been built in the year and a half since the commencement of
        construction in January 1974.

C271   "Szechwan Province Speeds Up Development of Gas Fields."
        Economic Reporter (English supp.) 3 (July/Sept. 1975):34-35.
        illus.
            A summary of the achievements made in the development of
        Szechuan gas field since 1965, including a 3.3-fold increase
        in production. Natural gas has become a major energy resource
        for the electric power, machine building, and salt refining
        industries. Soon it will be used as fuel for the production
        of sulphur in Szechuan province.

C272   "China Follows Her Own Path in Developing Industry--Taching,
        A Striking Example," by Chi Chin. China Pictorial 9 (Sept.
        1975):6-15. illus. (some color).
            A profusely illustrated story on the growth of the
        Tach'ing oil field during the past 15 years. Tach'ing has
        been developed through the spirit of self-reliance and without
        dependence on foreign assistance. Production of crude there
        has increased at an annual rate of 30%. At Tach'ing, China
        has created an integrated industrial and agricultural base.

C273   "China in the Big League." Petroleum Economist 42, no. 9
        (Sept. 1975):229-331. table.
            Summarizes a working paper presented as a background sur-
        vey at a conference on China's oil and trading possibilities
        held at Glasgow, Scotland, in June 1975 by the Sino-British
        Trade Council. Presents a glowing picture of China's rapidly
        expanding petroleum industry, with crude oil production jump-
        ing from about 8 million tons in 1965 to an estimated 50 mil-
        lion tons in 1973. The paper describes China's rapidly grow-
        ing export of crude oil and her major expansion of refining
        and transport facilities. As a result, China presents British
        industry with good opportunities for the sale of advanced
        technology and petrochemical facilities.

C274   "Industry Profile: Oil." Business China (Sept. 5, 1975):41.
        map.
            A report on oil production in China, discussing reserves,
        major fields, refining facilities, and trade.

C275   "U.S. Is Said to Bar Drilling Off China." <u>Wall Street Journal</u>
(Sept. 5, 1975):3.
    A news report quoting an article published in the Fall
1975 issue of <u>Foreign Policy</u> (entry C280) as saying that the
U.S. State Department has prevented two American firms, Gulf
Oil and Superior Oil, from drilling waters off mainland China
in order to avoid a possible boundary dispute between Taiwan
and China.

C276   "China's Oil: Oil under Disputed Waters." <u>Economist</u> 256
(Sept. 13, 1975):71. map.
    Discusses China's claims to oil deposits on the conti-
nental shelf that extend beyond her offshore territorial
limits. These claims have created a possible conflict with
Taiwan, South Korea, and the U.S., since some American oil
companies are drilling in the disputed waters under conces-
sions from these two countries. The U.S. State Department has
advised the American oil firms to stay clear of waters claimed
by China.

C277   "China's Bootstrap in Oil Production." <u>Forbes</u> 116, no. 6
(Sept. 15, 1975):39-41.
    A news analysis and commentary on China's oil production
and potential. Through her own efforts, China has been able
to produce up to 1.4 million barrels of crude a day. The
further expansion of oil production, however, requires technological progress and the tapping of offshore deposits.
China still seems reluctant to import the needed foreign know-
how to expedite her oil development.

C278   "Cartel's New Competitor," by David Pauly et al. <u>Newsweek</u> 86,
no. 13 (Sept. 29, 1975):68. illus., map.
    Discusses the possibility of China's becoming a competitor
to the OPEC oil cartel. Since China has large coal reserves,
she could use coal for the bulk of her internal needs and export surplus oil to earn funds necessary for industrialization. By 1980 China might be able to export 1/4 of her crude
oil output--950,000 b/d. This quantity, however, might not
greatly affect the price set by the OPEC nations.

C279   "Politics of China's Oil Weapon--China: The Next Oil Giant,"
by Jerome A. Cohen and Choon-ho Park. <u>Foreign Policy</u> 20 (Fall
1975):28-49.
    A worthwhile study of how China can use its growing oil
production to modernize her economy at home and increase her
clout abroad. Domestically, oil can be used to build up its
agriculture, petrochemical industry, and transportation;
internationally oil can be used as a tool to win friends and
gain influence in world politics. All these assumptions depend to a large degree on whether China can rapidly develop
her oil resources and produce in quantities comparable to
those of the Middle East or the Persian Gulf.

C280   "Time Bomb in East Asia--China: The Next Oil Giant," by Selig
       S. Harrison. Foreign Policy 20 (Fall 1975):3-27.
            An audacious and challenging hypothesis, assuming that
       China's petroleum output will reach 412 million tons (equiva-
       lent to Saudi Arabia's in 1977) by the year 1988, if offshore
       oil is fully exploited. Such growth would reduce global de-
       pendence on Middle Eastern oil and benefit oil-hungry coun-
       tries like Japan. However, Mr. Harrison contends that the
       development of China's offshore oil could touch off boundary
       disputes with Japan, South Korea, South Vietnam, and Taiwan,
       and that the U.S. had better avoid becoming entangled. This
       is a condensed form of the author's China, Oil, and Asia:
       Conflict Ahead?, published in 1977 (entry B39).

C281   "China: Three Major Oil Fields Reported." Sea Trade
       (Colchester, England) 5, no. 10 (Oct. 1975):89-91.
            Three new oil fields, so far unknown to the West, have
       been revealed: Wuchi, Chiling, and Chiuyisan, all believed
       to be located in Kwangtung province.

C282   "China's Production May Jump Six-Fold by 1985." World Oil 181
       (Oct. 1975):145-54. illus. (some color), maps.
            Discusses the background that led to China's emergence as
       a factor in Asia's oil trade. Also evaluates the potentials
       of China's offshore oil reserves and for increases in her oil
       exports, which will pay for imports of advanced technology.
       World Oil estimates China's offshore deposits at 33 billion
       barrels and her potential export volumes at 400,000 b/d to
       1 million b/d in 1980s.

C283   "Lure of the Offshore." Petroleum Economist 10 (Oct. 1975):
       391.
            Comments on Selig Harrison's "Time Bomb in East Asia"
       (entry C280). With recoverable offshore oil reserves esti-
       mated at 32 billion barrels, the Economist reasons, China has
       every incentive to clarify her legal and political disputes
       with other countries in order to establish a solid offshore
       position in her continental shelf, thus permitting a rapid
       growth in crude oil production.

C284   "Japan Seeks Longer-Term Oil." Business China (Oct. 3, 1975):
       46.
            Discusses China's unwillingness to provide long-term con-
       tractual arrangements with the Japanese for the sale of crude
       oil, a puzzle to experts in Japan's oil industry.

C285   "Canton Gears Up for Oil," by Paul Strauss. Far Eastern Eco-
       nomic Review 90, no. 45 (Nov. 7, 1975):32-34. illus.
            In the wake of the reported discovery of oil in Nanhai
       county, Kwangtung province, and the construction of a petro-
       chemical complex at the port of Whampoa, Strauss discusses the

petroleum situation of China in general and of Kwangtung in particular. He maintains that the latter could become a major oil base and an industrial center for petrochemical products if the Nanhai offshore oil fields prove to contain substantial reserves and come into full production.

C286   "Peking's Search for Technology," by Harold Munthe-Kaas. Far Eastern Economic Review 90 (Nov. 7, 1975):32-33. illus.
       A visit of China's nine-man oil delegation to Norway leads to speculation on the possibility of some limited cooperation between the two countries in exploring China's offshore oil. Norway's expertise in offshore technology, such as building drilling platforms and rigs, and offshore logistics, including undersea pipelines, is the focus of the Chinese visitors' attention.

C287   "Mainland Chinese Aim High with Big Petrochemical Plants." Oil and Gas Journal 73 (Nov. 10, 1975):197-98. illus., maps.
       Describes a large-scale petrochemical complex under construction at Liaoyang in Manchuria. The project costs $600 million and is designed to produce 87,000 tons a year of polyester and 46,000 tons of nylon, using primarily oil and gas from the Tach'ing oil field. Plant design and engineering are being handled by the French and construction by a Chinese army of 20,000 men under the authority of the Ministry of Light Industry. About 230 Chinese technicians are undergoing training in France. The completion of the complex is slated for 1978.

C288   "Snags Facing China's Oil Exports," by Susumu Awanohara. Far Eastern Economic Review 90, no. 48 (Nov. 28, 1975):42-43. illus.
       Reports the inferior quality of Tach'ing crude oil has caused problems too big for Japan's oil refineries to cope with.

C289   "China's Petroleum Industry," by Sreedhar. India Quarterly 31, no. 4 (Oct./Dec. 1975):382-85.
       Assays China's oil exploration effort, her crude oil output, domestic consumption, and export potentials. Sreedhar sees China's oil industry at the take-off stage, with vast potentials for exports by the year 1980 if it can obtain advanced technology and equipment from abroad.

C290   "China's Petroleum Industry Strides Ahead," by Min Ren. Economic Reporter (English supp.) 4 (Oct./Dec. 1975):41-43. illus.
       A historical review of China's oil industry, with emphasis on the development of China's three major oil fields: Tach'ing, Shengli, and Takang.

C291   "China's Chemicals," by Sy Yuan. <u>U.S.-China Business Review</u>
        2, no. 6 (Nov./Dec. 1975):37-53. illus., tables.
            One of the most informative reports on the state of
        China's chemical and petrochemical industry in 1975. The
        author traces chemical manufacturing from its primitive state
        in 1915 to the phase of rapid expansion into petrochemicals
        in the 1960s and then to its continued growth in the 1970s. He
        regards the import of advanced technology and plants from the
        West as vital to the future advancement of China's petrochemi-
        cal industry. Included in this report are excerpts of arti-
        cles on petrochemical works (from Chinese news media) in
        Peking and Shanghai, lists of some existing chemical plants
        operating in China and chemical plants acquired from abroad
        in 1972-1975, a table on characteristics of Chinese oil,
        statistics on imports and exports of petrochemical products,
        and a brief description of the Nanking Petrochemical Works.

C292   "Petrochemical Works in Peking and Shanghai." <u>U.S.-China
        Business Review</u> 6 (Nov./Dec. 1975):39.
            A translation of articles excerpted from Chinese media
        reports describing the operation of the Peking General Petro-
        chemical Works and the construction of the vast Shanghai
        General Petrochemical Works now under way in Chinshan county on
        the outskirts of the city of Shanghai.

C293   "A Visit to the Nanking Petrochemical Works." <u>U.S.-China
        Business Review</u> 6 (Nov./Dec. 1975):52-53. illus.
            An account of a visit to the petrochemical complex by a
        group of traders from Hong Kong in September 1974. The arti-
        cle provides a glimpse of the state of the petrochemical in-
        dustry in China, the operations of Chinese chemical plants,
        and Chinese workers' hopes and attitudes toward the current
        and future development of the industry.

C294   "The Sino-Japanese-Korean Sea Resources Controversy and the
        Hypothesis of a 200-Mile Economic Zone," by Choon-ho Park.
        <u>Harvard International Law Journal</u> 16 (Winter 1975):27-46.
            Analyzes territorial delineation and disputes over China's
        continental shelf in the Yellow Sea between China, Japan, and
        South Korea, each of which claims sovereign rights over poten-
        tially oil-rich offshore areas.

C295   "China's Strategy as an Oil Giant," by Joseph Kraft. <u>Los
        Angeles Times</u> (Dec. 16, 1975):sect. 2, p. 11.
            Predicts that the production of oil in China will be
        closely tied to her economic growth, with petroleum becoming
        a significant factor in China's total energy supply. The
        growing domestic demand for oil due to agricultural mechaniza-
        tion and industrial modernization requires petroleum as an
        energy source in the 1980s. Therefore, unless there is a

rapid expansion of crude oil output, China will not likely become an important exporter of oil in the 1980s.

C296   "The Houston-Peking Axis: Oil for China's Lamps--and More."
        U.S. News and World Report 79 (Dec. 22, 1975):27. illus.
        Houston is fast becoming an export center for oil tech-
        nology and equipment to China. Since Peking wishes to become
        a major oil producing power, its prime interest is in acquir-
        ing Western technology to accelerate its oil production.
        Houston is a base where the Chinese can purchase drilling
        rigs, exploration and production equipment and support facili-
        ties, pipelines, refineries, and petrochemical plants. Mil-
        lions of dollars worth of contracts have already been signed,
        including a $200 million deal with Pullman Kellogg Co. for
        eight fertilizer complexes.

C297   "Oil in China: Reserves, Production and Export Potential,"
        by P.K.S. Namboodiri. Journal of the Institute for Defence
        Studies and Analysis (New Delhi) 8, nos. 2-3 (Oct./Dec. 1975-
        Jan./Mar. 1976):283-307.
        Estimates China's oil reserves, both onshore and offshore,
        to be in the vicinity of 70-90 billion tons. With average
        annual gain of 20%, China may produce 130 million tons of
        crude oil by 1980, which would not make China a major oil
        exporting country. Rather, China will be a modest oil ex-
        porter in Sino-Japanese trade.

C298   "Visit to an Oil City." China Reconstructs 25, no. 1 (Jan.
        1976):12-19. illus.
        A report on the progress and achievements of the Peking
        General Petrochemical Works since its completion in 1969. The
        petrochemical complex is the largest of its kind in China and
        receives oil and gas through a 1,500-kilometer pipeline from
        the Tach'ing oil field. It has 8 major works comprising 18
        petrochemical plants, which turn out more than 50 different
        varieties of petrochemical products ranging from jet fuel and
        gasoline to synthetic rubber.

C299   "CIA Discounts China's Oil Role," by Theodore Shabad. New
        York Times (Jan. 18, 1976):45.
        A study released by the CIA on oil production prospects
        (entry B22) provides a less than optimistic picture of China
        as a future oil exporter. Rising domestic demand coupled with
        constraints in oil technology make China an unlikely oil giant
        in the foreseeable future. Accordingly, her oil export poten-
        tial is somewhat limited.

C300 "Major Chinese Oil Export by 1985 Doubted." <u>Oil and Gas Journal</u> 74 (Jan. 19, 1976):33. illus.
   Reports on First National City Bank's study of China's oil trade and forecast for her crude oil output and export potential. The bank predicts that China's oil exports to Japan and the Philippines will increase as the result of the steep rise in the price of crude since the 1973 oil embargo.

C301 "China Emerging as an Oil Power," by Fox Butterfield. <u>New York Times</u> (Jan. 25, 1976):sect. 3, p. 38. map, table.
   Analyzes China's emergence as a potential major petroleum producer and exporter. Compares Japanese and CIA estimates (see entry B22) of China's oil potential in the 1980s. The main constraint on China's becoming a major oil exporter is the growth in domestic demand for oil as a result of her industrialization and agricultural mechanization programs.

C302 "China-Korea Friendship Oil Pipeline." <u>Peking Review</u> 19, no. 5 (Jan. 30, 1976):24.
   This is a report on the completion of a crude oil pipeline connecting China and North Korea. This pipeline was presumably a joint Sino-Korean project, with each country building that segment lying within its territory. The opening of this pipeline means a direct flow of Tach'ing oil from northern Manchuria to oil refineries in North Korea.

C303 "China: A Bid for the Future," by Paul Fan. <u>Offshore</u> 35, no. 15 (Feb. 1976):71-73. illus., map.
   Evaluates China's onshore and offshore petroleum deposits in the areas of the north China basin and the Yellow, East China, and South China seas.

C304 "China Opts for Small Scale Energy Techniques," by Vaclav Smil. <u>Energy International</u> 13, no. 2 (Feb. 1976):17-18. illus.
   Examines the Chinese application of small-scale bio-gas as a substitute for natural gas in cooking. China has built hundreds of thousands of digesters made of decomposed plants and manure to generate bio-gas an an alternate fuel.

C305 "China to Become a Major Exporter?" <u>Petroleum Economist</u> 43, no. 2 (Feb. 1976):61.
   Speculates and comments on the prospects for China's becoming a major oil exporter. With the CIA's report (entry B22) discounting China as a prospective major exporter, and with Chinese crude oil having a very waxy content, the <u>Petroleum Economist</u> is not sure that China can export large volumes of oil in the future. Since Japan is keen on importing Chinese crude as a means of diversifying her source of supply, China is expected to provide Japan with 15-50 million tons a year during next 5 years. Other neighboring countries

such as North Korea, the Philippines, Thailand, Hong Kong, and
Vietnam are also regarded as prospective buyers of Chinese
oil.

C306   "Energy in the P.R.C.," by Vaclav Smil. Current Scene 13,
        no. 2 (Feb. 1976):1-16.
            An extensive survey and analysis of the development of
        China's energy resources (coal, electricity, gas, and oil).
        Smil evaluates deposits, production capabilities, technology,
        and export prospects.

C307   "'Historic Territorial Rights': Key to China's Oil Future,"
        by Donald Bakke. Offshore 35, no. 15 (Feb. 1976):73-76.
        illus., map.
            Analyzes China's territorial claims in the South China
        Sea, a potentially oil-rich area, which are in dispute with
        the Philippines, Malaysia, Vietnam, Indonesia, and Taiwan.
        The controversial zones include the Paracels, Senkaku, and
        Hsisha islands.

C308   "The PRC as an Oil Exporter," by Wan-li Wu. Southeast Asian
        Spectrum 4, no. 2 (Jan./Mar. 1976):1-8.
            Discusses China's oil reserves, exploration, and produc-
        tion. Estimates China's current volume of oil exports and the
        prospects for major shipments to Japan.

C309   "A Thriving Harbour--Chinhuangtao," by Chi Wen. China's
        Foreign Trade (Peking) 1 (Jan./Mar. 1976):28-29. illus.
            Reports on the completion of a trunk pipeline to the port
        of Ch'inhuangtao and describes the port as a major crude oil
        terminal for the shipment of oil to other parts of China by
        sea.

C310   "China's Self-Sufficiency in Petroleum: Problems and Pros-
        pects," by W.A.C. Adie. Foreign Affairs Reports 3 (Mar.
        1976):31-52. tables.
            Adie presents a comprehensive review and analysis of
        China's petroleum deposits, exploration, production, refining,
        transportation, and oil technology. China seems likely to be-
        come a leading oil producer, if she can overcome technological
        bottlenecks by importing foreign technology and receiving as-
        sistance from the West.

C311   "Energy in China: Achievements and Prospects," by Vaclav
        Smil. China Quarterly 65 (Mar. 1976):54-81. illus., map.
            Smil's China's Energy: Achievements, Problems, Prospects
        (entry B29) is summarized within this article. He covers
        China's development of coal, oil, gas, and electricity since
        1949 and concludes that China's energy resources potential is
        strong; the pattern of usage still backward; energy only mod-
        erately advanced; the aggregate output strongly expanding; and

the oil and gas industry particularly promising, with substantial crude oil export potentials. He predicts that the stage is set for a high level of energy production, especially oil and gas, if China can commit herself to orderly economic growth.

C312   "Japanese and China's Oil--Proceeding with Caution," by Alistair Wrightman. U.S.-China Business Review 3, no. 2 (Mar./Apr. 1976):31-35. illus.
        Reports from Tokyo on the difficulties and problems encountered in importing crude oil from China. Stresses Japan's need for planning a long-term accord on importation of Chinese oil, the extra cost incurred by Japanese refineries in processing Chinese crude because of its high paraffin content, and the increased costs of transportation because of China's inadequate port facilities to accommodate large-size tankers. The report calls for closer cooperation between Chinese and Japanese governments to iron out solutions.

C313   "A Sino-Japanese Crude Oil Contract, 1974." U.S.-China Business Review 3, no. 2 (Mar./Apr. 1976):36-37.
        A translation of the 1974 document on the accord between China's National Chemical Import and Export Corp. and Japanese importers. The terms of the agreement include product names, quantity, price, deliveries, delivery port, destination, payment, loading notice and conditions, loading regulations, product inspection, force majeure, and penalty. The original text of the contract was written and prepared in both Chinese and Japanese.

C314   "A New Petrochemical Complex Being Speedily Built." China Pictorial 4 (Apr. 1976):14-17. illus. (some color).
        Describes the construction of a large petrochemical complex under way on the outskirts of Shanghai in Chinshan county. The largest industrial project in Shanghai since 1949, Chinshan's first phase includes 6 plants and 4 auxiliary ones for the production of synthetic rubber and plastics. Currently, more than 50,000 workers are laboring day and night to complete the first stage of the project, which started in 1974.

C315   "China's Future as an Oil Exporter," by Chu-yuan Cheng. New York Times (Apr. 4, 1976):F14. illus., map.
        Presents a sobering analysis of the export potential of China's crude oil. Despite rapid increases in crude oil production between 1960 and 1974, China's oil output will tend to grow more slowly in the future, while domestic demand will rapidly accelerate. Oil available for export, therefore, will not amount to more than 25% of her total production. Consequently, China is unlikely to become a giant energy exporter, and her oil exports will probably not exceed 1.1 million b/d by 1980 and 2.5 million b/d by 1985.

C316   "Petroleum Output Increases." <u>Economic Reporter</u> (English
        supp.) 2 (Apr./June 1976):7.
            A summary of achievements in the Chinese petroleum indus-
        try in the first three months of 1976. Compared with the
        record year of 1975, crude oil output went up by 12.7%, while
        prospecting, oil field construction, and the production of
        petrochemical products all set new records.

C317   "A 6,011-Meter Deep Well." <u>Peking Review</u> 19, no. 19 (May 7,
        1976):30.
            Reports on success in drilling China's first deep well to
        a depth of 6,011 meters in Szechuan province. The drilling of
        such a depth means the penetration of sedimentary rock strata
        in southwestern part of the province, making possible the
        collection of gas and oil reserves in deep formations and
        facilitating the development of hydrocarbons in the Szechuan
        basin.

C318   "The Oil Enigma." <u>China Trade Report</u> 14 (June 1976):4.
            Analyzes China's contradictory and often enigmatic oil
        policy toward Japan. Drawing on Japanese news sources, the
        <u>Report</u> describes China's strange attitude during Sino-Japanese
        oil import negotiations in 1976. Initially, China seemed to
        be offended by a Japanese proposal for a lower import volume,
        only to inform Japan later of her intention to cut back crude
        oil exports even further. Apparently, this contradiction
        reflects the internal feud on oil export policy between the
        "Gang of Four" and their opponents.

C319   "Political Implications of the Petroleum Industry in China,"
        by Jessica Leatrice Wolfe. <u>Asian Survey</u> 16, no. 6 (June
        1976):525-39.
            Analyzes why political priority is accorded to China's
        oil industry over other sectors. Foreign policy and economic
        development may be the two major factors that have propelled
        the petroleum industry to its eminence in China.

C320   "Sino-Soviet Relations and Politics of Oil," by Arthur J.
        Klinghoffer. <u>Asian Survey</u> 16, no. 6 (June 1976):540-52.
            Theorizes that the transformation of China from her depen-
        dence on imported Soviet oil into an oil exporting rival in
        the East Asian market signifies a new turn in the relationship
        between China and the Soviet Union.

C321   "Large Oil Terminal Is Opened in China," by Theodore Shabad.
        <u>New York Times</u> (June 9, 1976):53, 63.
            Reports that the completion of an oil terminal at the port
        of Talien in southern Manchuria enables crude oil produced at
        Tach'ing to be transported by pipeline and then shipped
        abroad. Since the port can accommodate 100,000-ton tankers,
        China is in a better position than the Soviet Union to export

oil to Japan. The opening of the oil terminal also coincides with China's stepped-up efforts to develop her rich offshore oil deposit in the Pohai Gulf.

C322   "Rapid Progress in Marine Geological Survey." Peking Review 19, no. 24 (June 11, 1976):31-32. illus.
Reports on the technological advancement and geographical expansion of China's marine geological survey. China's effort to develop her undersea oil deposits is evidenced by the building of her first marine geological drilling vessel, the K'ant'an No. 1, and her recent expansion of offshore activities from the Pohai Gulf, an inland area, to the Yellow, East China, and South China seas. A Chinese-made offshore rig was reported to have struck oil in the southern part of the Yellow Sea.

C323   "China: Ideology, Then, Oil," by Peter Weintraub. Far Eastern Economic Review 92, no. 25 (June 18, 1976):36-38. illus.
The ouster of Deputy Premier Teng Hsiao-p'ing has cast a damper on immediate prospects for China's becoming a major crude oil exporter.

C324   "Builders of the Taching Oil Field." China Reconstructs 25, no. 7 (July 1976):32-35. illus.
An account of interviews with two oil field workers: Wu Chuan-ching, an oil driller, and Sun Hai-chen, a female pipe fitter. It also reports on the community life surrounding the Tach'ing oil field and, in particular, describes the working housewives who tend farms in the agricultural settlements.

C325   "China's First Oil Drilling Ship." China Reconstructs 25, no. 7 (July 1976):36-40. color illus.
Describes the building of China's first drilling vessel, Pohai No. 1. It is a jackup platform for shallow-water drilling, carrying machinery, equipment, and quarters for oil workers. The vessel was initially designed in 1961, but construction did not start until 1970 on account of numerous technical and other problems. Pohai No. 1 was finally commissioned in 1973 and has been operating smoothly ever since.

C326   "New Oil Port in Talien." Peking Review 19, no. 27 (July 2, 1976):24. illus.
A news report on the completion of China's largest oil terminal at the port of Talien. The crude oil wharf has two deep-water berths capable of accommodating tankers of 100,000 tons and 50,000 tons each. It is equipped with Chinese-made advanced facilities such as automatic loading arms and a 1,400-meter-long jetty. Construction commenced in 1974, and the whole project was completed within a year and half. This new oil port will greatly facilitate the transportation of crude oil from the Pohai Gulf to Japan and other coastal areas of eastern China.

C327   "China Driving to Boost Oil Production." Oil and Gas Journal
        74 (July 5, 1976):42-43.
            Since 1970, oil production has been boosted 300% to 1.65
        million b/d. Both exploration and development programs have
        been accelerated. Refining capacity in 1975 is estimated to
        be at 1.228 b/d, or 80% of China's crude oil output of 1.534
        million b/d.

C328   "China Builds Own Rigs for Pohai Search." Oil and Gas Journal
        74 (July 19, 1976):58-59. illus.
            Describes the Chinese-made jackup rig Pohai No. 1 during
        drilling in shallow waters of the Pohai Gulf. The rig was
        built by the Hungch'i Shipyard at the port of Talien. Japa-
        nese sources estimate the Pohai Gulf oil deposits to be in the
        range of 12-15 billion tons, of which half are considered
        recoverable.

C329   "China's First Ultra-Deep Oil Well." China Reconstructs 25,
        no. 8 (Aug. 1976):20-21. illus.
            By using both modern and indigenous methods, and by ele-
        vating ideological consciousness, Drill Team No. 7002 in
        Szechuan province has succeeded in drilling an oil well 6,011
        meters deep. This is the first time that a drill has been
        able to penetrate the sedimentary strata of central Szechuan.

C330   "Chinese Oil Production Slows but Long-Term Prospects Look
        Good," by Vaclav Smil. Energy International 13, no. 8 (Aug.
        1976):25-26. illus.
            The slowdown in China's export of oil to Japan from a peak
        in 1974 to somewhat smaller volumes in 1975 leads the author
        to conclude that bottlenecks in transportation facilities
        (e.g., inadequate pipeline capacity and port facilities) are
        hampering oil exports.

C331   "Mowming--Shale Oil City." China Reconstructs 25, no. 8
        (Aug. 1976):18-19. illus.
            A sketch of the city of Mowming (or Maoming) in Kwangtung
        province as a growing oil city of southern China. The ex-
        ploitation of shale oil started in 1958 and since then this
        industrial city has built 7 refining units with 12 auxiliary
        plants. The production of shale oil in 1975 is 389% above
        1965, and crude oil output is up 422%.

C332   "Production at Takang Oilfield Returns to Normal after Earth-
        quake." Economic Reporter (English supp.) 3 (July/Sept.
        1976):13.
            Crude oil production and shipment of oil were kept un-
        interrupted despite the devastating earthquake that struck at
        T'angshan-Fengnan area on July 28, 1976. Oil production was
        restored to preearthquake levels the second day after the
        calamity.

C333   "China's Energy Resources," by Chu-yuan Cheng. <u>Current History</u> 71, no. 419 (Sept. 1976):73-75, 84. tables.
       A general survey of China's energy situation as of mid-1976. Cheng first traces the development of China's hydro-carbon resources. He asserts that oil and gas have come to account for nearly 32% of total Chinese energy production as opposed to a mere 3% in 1957 and 14% in 1965. He observes that the opening of major oil fields in north and northeast China has brought about changes in the regional distribution of energy, with sources of supply now being closer to energy consumption centers. Cheng concludes that China's energy prospects (particularly the supply of oil and gas) are quite promising.

C334   "Communist China's Oil Exports Revisited," by Vaclav Smil. <u>Issues and Studies</u> 9 (Sept. 1976):68-73.
       Smil asserts that the latest development verifies his pes-simistic assessment of China's oil export potential that ap-peared in the March 1975 issue of this same journal. He stresses the higher costs of China's oil due to transportation bottlenecks and the poor quality of Tach'ing crude. He con-cludes that the amount of Chinese oil available for export will not exceed 46 million tons in 1980 and 62 million tons in 1985. This will probably account for only 1.5-3% of total world petroleum trade at that time.

C335   "Temporary Stagnation in Output." <u>Petroleum Economist</u> 43, no. 9 (Sept. 1976):358.
       A sharp decline in the rate of growth of China's crude oil production leads the <u>Economist</u> to speculate on the possible causes for such a big drop from previous growth of more than 20% a year. The <u>Economist</u> maintains that the current slowdown could be caused by a slower expansion in refinery capacity and possible bottlenecks in transportation and export markets. China's internal political strife might also be a factor.

C336   "One Firm's Experience in Selling a Large Chemical Plant to China." <u>Business China</u> (Sept. 3, 1976):41-43. illus.
       Describes a practical experience of negotiating, selling, and building a huge petrochemical complex in China worth half a billion dollars. When Technip, a French concern, took on the job, it was the biggest oil-related project China had ever entered into with a foreign firm. The complex was built at Liaoyang between 1971 and 1976. The contract, signed in 1973, included the planning, engineering, equipment, supply, and technical assistance for a 21-unit petrochemical facility capable of producing 87,000 tons of polyester and 46,000 tons of nylon per year.

C337   "China's Oil Policy," by King C. Chen. Yale Review 56
        (Autumn 1976):1-13.
            Assesses the political and economic implications of
        China's oil policy toward her neighboring countries. An oil
        deal with Japan will politically undermine the latter's nego-
        tiations with the Soviet Union over the development of Sibe-
        rian energy and enhance the possibility of a Sino-Japanese
        peace treaty. Chinese exports to ASEAN countries will serve
        as an economic and psychological counterpart to Japanese
        influence in Southeast Asia. To the Third World, Chinese oil
        means endorsing its independent possession of natural re-
        sources. From an economic standpoint, oil can be used to
        acquire technology for China's industrial modernization.

C338   "China's Oil Trade in the 1980s: A Closer Look," by Sidney
        Klein. Military Review 56, no. 10 (Oct. 1976):52-55.
            Rebuts the Kraft-CIA estimates of Chinese oil's export
        potentials (see entry B22), which are described too conserva-
        tively. Klein maintains that China at least should become a
        major oil exporter in Asia by the 1980s, when 10% of her crude
        oil production of over 400 million tons should become avail-
        able for export. The author further contends that Chinese oil
        should be a significant factor in the Asian oil market in the
        1980s.

C339   "New Deep-Water Berths: Oil Pier at Huangtao." China
        Reconstructs 25, no. 10 (Oct. 1976):42-46. illus., map.
            Depicts the building and expansion of China's major deep-
        water oil ports such as Ch'inghuangtao, a key oil terminal in
        Pohai Gulf; Talien, the largest oil port in China; and the
        newly-built port of Huangtao located in Chiaochou Bay in
        Shantung peninsula.

C340   "Slowing Down or Dying--The Oil Boom--China '76," by Peter
        Weintraub. Far Eastern Economic Review 94, no. 40 (Oct. 1,
        1976):55-56.
            The slowdown in crude oil output in the first half of 1976
        leads the author to speculate on the possible causes and con-
        sequences. He maintains that China's leftist leadership has
        been critical of Teng Hsiao-p'ing's economic line, which
        places oil as a key to China's overall economic development.
        The ascendency of the leftist group might explain the slower
        growth in China's oil production. Beside that, there are
        other factors that caused the decrease in China's oil exports,
        such as the waxy content of its crude and lower Japanese
        demand.

C341   "Assessing China's Oil Industry." China Trade Report 14
(Nov. 1976):6-8.
    Evaluates China's petroleum industry as of 1976. It
attempts to estimate China's oil deposits; to discern and
delineate her oil policies; to forecast short-term and medium-
term production and export prospects; and to predict the role
of oil in China's overall foreign trade. Much of the analysis
centers on a variety of internal and external factors that
affect China's decision-making and the formulation of her
energy and oil policy.

C342   "Mainland China's Oil Industry: A Study of Its Development,"
by Chün Chang. Issues and Studies 11 (Nov. 1976):74-91.
tables.
    A sober and soul-searching appraisal of China's oil indus-
try up to 1976. The author critically examines China's reserves and its history of growth. He considers China's technical expertise in oil to be very backward and her potential
to export very limited. Eight estimates of China's oil output
are listed and compared. The author regards the prospects of
China's petroleum industry as dim, since China seems unable to
overcome deficiencies in equipment, manpower, capital, transportation capacity, and above all, modern expertise.

C343   "Growth Seen in China's Oil Equipment Buying." Oil and Gas
Journal 74 (Nov. 8, 1976):148.
    Ray Pace, President of Baker Trading Co., says that China
will resume her overseas purchase of petroleum equipment after
mid-1977. China has an estimated 600 land rigs, of which 114
were bought from Rumania. Several seismograph systems and
sophisticated data processing units have been purchased from
American firms. The Chinese seem to be testing advanced
equipment made in the U.S. and should be placing orders for
them in 1977.

C344   "Two Chemical Installations Go into Operation in China."
Petroleum Times 80, no. 2040 (Nov. 26, 1976):21.
    Two of the eight agricultural-chemical complexes, each
producing 1620 metric tons of urea fertilizer a day, have gone
into operation in China. Designed by a Houston-based sub-
sidiary of Pullman Kellogg Continental of Amsterdam, the
plants, built in Liaoning and Heilungkiang provinces, have
since gone on line.

C345   "Council Visits Taching." U.S.-China Business Review 6
(Nov./Dec. 1976):43. illus.
    An account of visits to the Tach'ing facilities by a dele-
gation of the National Council for U.S.-China Trade. It reports on the oil field, the production average of 70-100 tons
per well a day, its extracting method--water injection into
the oil bearing strata--and the production of gasoline, jet

fuel, kerosene, and diesel fuel. The delegation also toured
the Tach'ing Oil Refinery, which has a 100,000 b/d refining
capacity.

C346 "Earthquake Strikes at China's Energy Centers," by Vaclav
Smil. Energy International 13, no. 12 (Dec. 1976):21-22.
illus., map.
The devastating earthquake that struck the T'angshan area
of north China on July 28, 1976 may have also damaged China's
major energy-producing facilities there. In addition to the
possible destruction of a major coal mine, varying degrees of
damage occurred at the petrochemical complex in Tientsin, the
Takang oil fields nearby, an oil pipeline, and at the oil
terminal at Ch'inhuangtao.

C347 "Paleo-Oceanic Crust of the Chilienshan Region, Western China,
and its Tectonic Significance," by W. Quan and L. Xueya.
Scientia Geologica Sinica 1 (Jan./Mar. 1976):42-55. Excerpted
in Deep-Sea Research and Oceanographic Abstracts 12 (Dec.
1976):721.
Discusses tectonic composition and oceanic crust in
Ch'ilienshan region, located in the northern part of the
Tsaidam basin. The area is rich in hydrocarbon deposits.

C348 "Petroleum Geology and Industry of the People's Republic of
China," by Arthur A. Meyerhoff and J.O. Willums. U.N. Economic and Social Commission for Asia and Pacific, Committee
for Coordination of Joint Prospecting for Mineral Resources
in Asian Offshore Areas. Technical Bulletin 10 (Dec. 1976):
103-212. illus., tables.
One of the most complete and detailed analyses and evaluations of China's onshore and offshore petroleum potentialities,
by two geologists of stature. This comprehensive survey also
updates and upgrades A.A. Meyerhoff's previous treatise entitled "Developments in Mainland China, 1949-1968" (entry
C99).

C349 "East Asian Coasts, Offshore Are Promising Petroleum Frontiers," by A.A. Meyerhoff. Oil and Gas Journal 74 (Dec. 27,
1976):215, 224. maps.
Meyerhoff, a celebrated geologist, evaluates and analyzes
possible oil deposits off the coast of China and other East
Asian countries. The Pacific continental shelf along the
Mekong delta, the South China Sea, and Chinan-Pohai-Liaoho,
are considered to be areas of great promise. The Pohai Gulf
area may have oil reserves of 30 billion barrels. China
therefore seems to possess adequate reserves to attract foreign capital capable of developing a large-scale offshore oil
industry.

C350   "Worldwide Report," by Larry Aulderidge. Oil and Gas Journal
       74 (Dec. 27, 1976):101-80.
            The Journal's 1976 annual listings of oil and gas outputs,
       oil and gas fields, and refinery capacities of major oil
       producing and consuming countries, including China and Taiwan.
       It also covers the world's major continental shelves and
       evaluates China's Chinan-Pohai-Liaoho area as most promising
       zones, possibly containing 30 billion barrels of oil reserves
       and deposit adequate to attract foreign capital to build a
       potentially large-scale offshore oil industry.

C351   "China's Oil Policy," by Jerome A. Cohen and Choon-ho Park.
       In Post-Mao China and U.S.-China Trade, edited by Shao-chuan
       Leng. Charlottesville, Va.: University Press of Virginia,
       for the Committee on Asian Studies, University of Virginia,
       1977, pp. 108-40.
            Cohen and Park update and modify their previous article,
       "Politics of China's Oil Weapon," published in 1975 (entry
       C279). In this essay, they assess developments that have
       occurred since and take a more critical view of adverse fac-
       tors that tend to limit China's prospects as a giant oil pro-
       ducer and exporter. They have revised somewhat downward
       China's output and reserve figures. The authors also examine
       China's domestic demand and export potential more thoroughly
       and conclude that in the foreseeable future, China will not
       become a major oil exporter, that most of her surplus oil will
       be exported to Japan, and that her political influence as an
       oil power will be limited.

C352   "Perspectives on Energy in the People's Republic of China,"
       by Vaclav Smil and Kim Woodard. In Annual Review of Energy
       2 (1977):307-42. tables.
            An extensive survey of China's energy situation in the
       year 1977, covering coal, electric power, gas, and petroleum.
       The authors offer fairly broad examinations of China's oil and
       gas industry, assessing reserves, exploration, production,
       refining, and consumption.

C353   "China's Sovereignty on Outer Continental Shelf is Inviola-
       ble." Peking Review 20, no. 25 (Jan. 17, 1977):16-17.
            Largely a translation of articles from Jen-min jih-pao
       (People's Daily) and statements by China's Ministry of Foreign
       Affairs protesting the infringement of Chinese sovereignty by
       the Japanese Diet when it approved an agreement with South
       Korea on the joint development of oil resources in the East
       China Sea.

C354 "Petroleum and Coal-Mining Industries' Achievements." <u>Peking Review</u> 20, no. 3 (Jan. 14, 1977):7.
China claims that the output of crude oil, natural gas, and other petrochemical products in 1976 have exceeded the state plan. Crude is up 13% over last year and natural gas 11%. It also states that many promising new oil fields have been developed last year.

C355 "Limits on China's Oil Exports," by Peter Weintraub. <u>Far Eastern Economic Review</u> 95 (Jan. 21, 1977):100-3. illus.
The <u>Review</u>'s veteran reporter analyzes two factors that constrain China's export potentials: the quality of Chinese crude oil and port facilities inadequate to accommodate super-tankers. The high paraffin content of Chinese crude requires costly modification of the refineries by the importing coun-tries. Also, the inability of Chinese ports to accommodate tankers larger than 100,000 tons means higher cost for trans-portation by sea. Although China seems to be charging less for crude exported to Thailand and the Philippines, the author contends that in the final analysis this might not be the best answer.

C356 "Taching Chemical Fertilizer Plant." <u>Peking Review</u> 20, no. 4 (Jan. 21, 1977):31.
A large-scale chemical fertilizer plant, using gas and by-products generated at the Tach'ing oil field as raw material, is completed ahead of schedule and goes on line. The plant has a capacity of 1 million tons of chemical fertilizer. Construction started in May 1974 with the assistance of 180-odd factories throughout China. Part of the facility is imported.

C357 "China's Oilfield." <u>Business Week</u>, no. 2645 (Jan. 24, 1977): 35. map.
Confirmation of the discovery of China's fourth major oil field (Huapei) in northeastern China prompts <u>Business Week</u> to speculate that China will boost crude oil exports to pay off her foreign debts. It suggests that China's new leadership may be looking for foreign buyers for its oil and that Japan seems to be the ideal candidate.

C358 "300,000-Ton Ethylene Project." <u>Peking Review</u> 20, no. 5 (Jan. 28, 1977):31. illus.
A large ethylene plant with a capacity of 300,000 tons a year is completed at Tach'ing. It took only two years and eight months to build this giant plant, which produces plas-tics, synthetic rubber, and other petrochemical products. The facilities were partially imported, but Chinese were responsible for supervision and construction.

C359   "China Regaining Former Oil Production Growth." <u>Petroleum
        Economist</u> 44, no. 2 (Feb. 1977):55.
            Analyzes China's announcement of a 13% gain in crude oil
        production in 1976 over 1975. Since previous annual rates of
        increase have been over 20%, China's recent announcement might
        mean a slowdown in her growth rate. The <u>Economist</u> assumes a
        slower pace of growth in the output of oil in the first half
        of 1976 and a faster pace during the second half of the year.
        Accordingly, it estimates China's crude oil output in 1976 to
        be 87 million tons, well on its way toward a 100 million-ton
        output in 1977.

C360   "New General Petrochemical Works." <u>Peking Review</u> 20, no. 6
        (Feb. 4, 1977):31. illus.
            A new general petrochemical plant, under construction
        since April 1966, has been put into operation in China. The
        plant is located at the Shengli oil field in Shantung prov-
        ince. It comprises an oil refinery, two chemical fertilizer
        plants, a catalyst plant, and a synthetic rubber plant, and
        produces a variety of more than thirty petrochemical products,
        including gasoline, kerosene, diesel oil, and tar.

C361   "Ten Major Contributions of Taching Oilfield." <u>Economic
        Reporter</u> (English supp.) 1 (Jan./Mar. 1977):15-16. illus.
            Lists major achievements made at the Tach'ing oil base in
        the past 17 years. Among the most remarkable gains are 15-
        fold returns on original investment in the oil field, a 28%
        annual growth rate in crude oil output, the building of a
        petrochemical complex, and the integration of agriculture and
        the oil industry at Tach'ing.

C362   "China and Its Oil," by Pierre de Villemarest. <u>Total Informa-
        tion</u> (Geneva) 69 (Spring 1977):2-11. color illus., map,
        tables.
            An extensive report on the state of China's petroleum
        industry based on information from France. It covers the
        geographical locations of China's oil fields (with special
        accounts on six major oil producing basins), oil production,
        major oil refineries, China's need for oil technology, the
        transportation of oil, and China's oil diplomacy. Also cov-
        ered are China's natural gas and shale oil industries.

C363   "China Short of Fuel." <u>Petroleum Economist</u> 44, no. 4 (Apr.
        1977):140.
            China's recent campaign to economize on fuel consumption
        has raised the question of whether she is suffering from a
        shortage of energy. The disruption of coal production caused
        by a severe earthquake at the Kailan coal complex and China's
        promise to deliver only 5.4-6.2 million tons of crude oil to
        Japan in 1977 provides food for thought. Previous forecasts
        of China's becoming a major oil exporter may be premature.

C364   "Japan: Mixing Oil and Steel," by Phijit Chong. China Trade
       Report 15 (Apr. 1977):4-5. illus.
           Discusses the dilemma Japan's industry faces when it
       imports costly oil from China in exchange for steel and other
       machinery. China's waxy oil has been troublesome for Japan's
       oil consuming industry--electric power companies. Removing
       the paraffin content requires expensive cracking facilities
       for Japan's refining industry. But in order to export steel,
       chemicals, and machinery to China, Japan has no alternative
       but to reciprocate--by importing Chinese oil.

C365   "Large Modern Petrochemical Base." China Reconstructs 26,
       no. 4 (Apr. 1977):49. illus.
           A report on the completion of the Shengli General Petro-
       chemical Works near the Shengli oil field in Shantung province.
       This petrochemical complex consists of an oil refinery, two
       fertilizer plants, one catalyst plant, and a synthetic plant.
       It produces gasoline, kerosene, diesel oil, asphalt, chemical
       fertilizers, acrylonitrile, and benzoids, using oil and gas
       from the Shengli oil field as raw materials. Construction of
       the complex started in 1966.

C366   "Relation of the Tectonics of Eastern China to the India-
       Eurasia Collision: Application of Slip-Line Field Theory to
       Large-Scale Continental Tectonics," by P. Molnar and
       P. Tapponnier. Geology 4 (Apr. 1977):212-16. illus.
           Focuses on the tectonics of northeast and southeast China,
       finding the former to be deformed, the latter rather stable.
       Both are seemingly the results of the tectonic collision of
       the India-Eurasia land mass.

C367   "Dutch, U.K., U.S. Firms Vie for Petroleum Equipment Sales."
       Business China (Apr. 2, 1976):19.
           In addition to American manufacturers, British and Dutch
       makers of offshore oil drilling facilities are visiting China
       in an attempt to sell equipment there.

C368   "Development of Petrochemical Industries in Republic of
       China," by Kwang-shih Chang. Industry of Free China 4
       (Apr. 25, 1977):2-8.
           A brief sketch of the phenomenal growth of Taiwan's petro-
       chemical industries from their tiny base in the 1950s to their
       present size. Chang discusses a variety of factors that con-
       tributed to their rapid development, including heavy government
       investments, the involvement of foreign interests, and the
       role of the private sector. The author sets forth five condi-
       tions as prerequisites for future growth: the availability of
       raw materials, the price of naphtha, technology and equipment
       fabrication, the availability of adequate infrastructure, and
       the availability of skilled manpower.

C369   "For More than Oil--Taching Impressions (I)," by Shan-hao
Chiang. <u>Peking Review</u> 20, no. 19 (May 6, 1977):40-45.
    The first of the author's four on-the-spot reports about
the Tach'ing oil base. Chiang describes the early history of
this oil field: an army of 10,000 workers, battling inclement
natural elements in the desolate prairie of northeast China,
opened up Tach'ing in 1959 and (using Mao's thought for
guidance) built it into an industrial-agricultural complex.

C370   "Taching's Theme Is Growth," by Peter Weintraub. <u>Far Eastern
Economic Review</u> 96, no. 18 (May 6, 1977):51-52.
    Reports on the proceedings of China's national industrial
conference held at Tach'ing. The national congress accents
orderly economic growth and the speedy development of China's
petroleum industry. In a 30,000-word keynote speech, the man
in charge of the Tach'ing oil field calls for tighter, more
centralized control of the field, improved management, and
imports of advanced foreign technology for the rapid development of China's oil industry.

C371   "Stand Up Straight--Taching Impressions (II)," by Shan-hao
Chiang. <u>Peking Review</u> 20, no. 20 (May 13, 1977):25-29.
illus.
    Emphasizes willpower and the fierce spirit of self-
reliance as driving forces in the building of the Tach'ing oil
field and its petrochemical complex in the 1960s. The oil
base is completely designed and constructed by the Chinese.
Ideology seems to take lead over expertise.

C372   "Black Gold and the Red Flag--Taching Impressions (III)," by
Shan-hao Chiang. <u>Peking Review</u> 20, no. 21 (May 20, 1977):
20-24.
    In this article, the author again emphasizes ideology as
the major motivating factor behind the building of the
Tach'ing oil base. To promote an esprit de corps, the mana-
gerial personnel join the workers in physical labor and share
the same housing.

C373   "Combining Urban and Rural Life--Taching Impressions (IV)," by
Shan-hao Chiang. <u>Peking Review</u> 20, no. 22 (May 27, 1977):
24-27.
    After more than two decades of growth, Tach'ing has become
a sprawling industrial-agricultural complex. It has an oil
field, a petrochemical complex, administrative centers, and
agri-business establishments formed by 60 villages comprising
164 rural settlements, with a total population of nearly half
a million people. In such an ideal environment, men work in
the oil fields and the factories, able-bodied women tend the
fields, and children walk to their nearby schools.

C374   "Chairman Hua Inspects Taching." Economic Reporter (English
        supp.) 2 (Apr./June 1977):16-21. illus.
            An account of Hua Kuo-feng's visit to the Tach'ing oil
        field on April 17-19, 1977. During his visit, Hua inspected
        oil wells, agricultural settlements, and the General Petro-
        chemical Works, including the Tach'ing Fertilizer Plant. This
        was Hua's first visit to the oil base since becoming Party
        Chairman in May 1976.

C375   "China's Biggest Oil Base--Taching Oilfield." Economic
        Reporter (English supp.) 2 (Apr./June 1977):29-33. illus.
            Thirteen color photographs of Tach'ing portray the skyline
        of the petrochemical works and workers in action in the oil
        fields and on the farms.

C376   "Taching Overfulfills Quarterly Production Plan." Economic
        Reporter (English supp.) 2 (Apr./June 1977):27.
            Despite encountering the most severe cold spell of the
        decade, workers at the Tach'ing oil field have overfulfilled
        crude oil quotas for the first quarter of this year. All-time
        records have been set for crude oil output, oil refining, vol-
        ume of water injection in an oil-bearing bed, and total value
        of industrial output.

C377   "Taching--Oilfield with Both Rural and Urban Characteristics."
        Economic Reporter (English supp.) 2 (Apr./June 1977):26-27.
        illus.
            A sketch of the agricultural settlements surrounding the
        Tach'ing oil field, focusing on a typical one called Chuangyeh
        (Pioneer) Village. It has 590 families with a total of 2,500
        residents, with 520 housewives doing farm work while others
        are given duties at the dining halls, sewing stations, or
        nurseries. This means that all able-bodied persons do some
        kind of work. About half a million oil base workers and their
        families live in 60 villages like Chuangyeh village and 163
        residential centers.

C378   "Taching--Red Banner on China's Industrial Front." Economic
        Reporter (English supp.) 2 (Apr./June 1977):24-25. illus.
            A brief review of the speedy development of the Tach'ing
        oil field. It gives a chronological description of how the
        oil base was built and how decisions were made on exploiting
        it. The Tach'ing petrochemical complex is characterized as
        "combining industry and agriculture, integrating town and
        country and thus set[ting] an example for building a new type
        of socialist enterprise in China."

C379   "Energy Resources of China," by Sreedhar. <u>China Report</u> 3
       (May/June 1977):17-22.
           This discussion and appraisal of China's energy resources
       emphasizes her coal and oil. The author, an Indian writer,
       includes a brief survey of China's oil reserves, major oil
       fields, crude oil production, refining capacities, and exports
       from 1949 to 1974.

C380   "China's Oil Prospects," by Bruce J. Esposito. <u>Military
       Review</u> 57, no. 6 (June 1977):14-20. illus. First published
       in <u>Asian Affairs</u> (July/Aug. 1976):15-20.
           Esposito assesses the export potential of Chinese oil and
       discusses the factors inhibiting China from becoming a major
       oil exporter. A steep increase in domestic oil consumption,
       inadequate port facilities and pipeline capacity, the poor
       quality of Chinese oil, and soft oil prices globally are re-
       garded as the major reasons. China's oil surplus for export,
       moreover, accounts for only 10-15% of her total crude oil
       production--a very modest amount in comparison with Saudi
       Arabia.

C381   "Taching today." <u>China Pictorial</u> 6 (June 1977):14-16. illus.
           The illustrated story of 17 years of consecutive advances
       in operating the Tach'ing oil base and in building its petro-
       chemical complex. Since its discovery and development,
       Tach'ing has returned to China's treasury 14 times the amount
       of its original investment.

C382   "Taching Fights the 'Four Pests.'" <u>Peking Review</u> 20, no. 24
       (June 10, 1977):17-20.
           An excerpt from the speech of the party boss and chairman
       of Revolutionary Committee at Tach'ing, Sung Chen-min, at the
       National Conference on Learning from Tach'ing in Industry. It
       discusses the damage done to the development of the oil base
       by the henchmen of the discredited "Gang of Four." Internal
       strife created by the Cultural Revolution has retarded
       Tach'ing's progress.

C383   "Aussie Trial Purchase of Chinese Crude Oil Raises Hopes for
       Sales." <u>Business China</u> 12 (June 17, 1977):80.
           The arrival of a 2,000-ton trial shipment of Chinese crude
       has sparked hopes for a greatly enhanced two-way trade between
       Australia and China.

C384   "Improved Outlook for Chinese Oil Output, Dollar Earnings,
       Buying." <u>Business China</u> 12 (June 17, 1977):77-80. map,
       tables.
           Discusses American, Japanese, and Swiss estimates of
       China's 1976 oil production. It also examines China's oil
       reserves and speculates on the volume of her crude oil exports
       by the year 1980.

C385   "China Expands Oil Search." <u>Washington Post</u> (June 27, 1977): A17.

        Reports on the massive oil hunt being stepped up in Szechuan province in response to the party's call for the building of ten or more of Tach'ing-type oil fields in next ten years.

C386   "New Course for Oil Policy," by B.A. Rahmer. <u>Petroleum Economist</u> 44, no. 7 (July 1977):259-60. map.

        An analysis of China's prospects in oil production under the new leadership of Premier Hua Kuo-feng. Under Hua's pragmatic leadership, China seems to be ridding itself of inertia and striving for industrial progress, as in its plans to build another ten or more Tach'ing-type oil fields. This would mean that China has set a target of producing around 500 million tons of crude oil by the end of this century. The author considers the chances to be good for China to reach this goal in view of its past performance.

C387   "Taiwan Chemicals Trade Boom as Export Sales Hit $8.2 Billion." <u>Chemical Marketing Reporter</u> (July 4, 1977):4, 12.

        Taiwan's petrochemical industry is growing at a rapid pace. Complete statistics indicate a 28% annual growth rate. Ethylene production more than doubled in 1976 and polyethylene output also hit the 1 million ton mark, for a 67% gain. A gain of 50% was scored in polyvinyl chloride and polystyrene. As a result, the export of petrochemical products is largely responsible for hiking total export volumes to $8.2 billion.

C388   "China's Future as a Leading Oil Producer Is Questioned in Study Released by CIA." <u>Wall Street Journal</u> (July 5, 1977):8.

        According to a study by the CIA (entry B40), the prospects of China's becoming a major oil producing nation do not seem bright. Because of backward oil technology, bottlenecks in transportation, inferior oil quality, and growing domestic demands, production will not match that of Saudi Arabia, and exports will not be significant in the foreseeable future. Unless China allows international oil companies to participate in the development of its petroleum resources, the study concludes, a major boost in production is unlikely any time soon.

C389   "Active Faulting and Tectonics in China," by P. Tapponnier and P. Molnar. <u>Journal of Geophysics</u> 20 (July 10, 1977):2905-30. illus.

        Explores China's tectonics and her geological faults system by interpreting satellite imagery augmented by seismic data. The locations of some of China's geological fault systems are identified.

C390   "Underground Oil Tank." Peking Review 20, no. 29 (July 15,
        1977):31.
            Reports that the first underground stone-cave oil tank,
        built into solid rock under water to store petroleum, has been
        completed. The merit of an underground oil tank is that it
        takes less time, less space, less steel to construct; and
        less maintenance. Investment cost is moderate, too.

C391   "CIA Doubts China to Be a Major Crude Exporter." Oil and Gas
        Journal 75, no. 29 (July 18, 1977):46-47.
            Reports on a special study on China's oil potential
        released by the CIA (entry B40). The study discounts China's
        prospects of ever becoming a major oil exporter. The CIA
        contends that China does not have adequate financial resources
        and technological knowhow to exploit her vast offshore oil
        deposits. It concludes that unless China allows a huge infu-
        sion of foreign investment and participation in her offshore
        ventures, she will be unable to produce substantial quantities
        of oil for export.

C392   "Record Oil Output." Peking Review 20, no. 30 (July 22, 1977):
        3-4.
            Crude oil production in the first half year of 1977 is
        10.6% above that of 1976. The drilling footage in May and
        June 1977 was double the average monthly footage in the first
        quarter of last year. Production at all major oil fields
        (Tach'ing, Shengli, Huapei, Yümen, and Karamai) have all
        topped their half-year quotas.

C393   "CIA's Cautious Assessment." Petroleum Economist 44, no. 8
        (Aug. 1977):320.
            Comments on the CIA's report (entry B40) about China's
        prospects as an oil producer and exporter. Some of the CIA's
        estimates seem to coincide with the Economist's own. Accord-
        ing to the CIA, China's onshore reserves are about 40 billion
        barrels, with her 1976 crude oil output estimated at 1.7 mil-
        lion b/d and 1980 projections at between 2.4 and 2.8 million
        b/d. The CIA also forecasts an exportable surplus of only
        200-600,000 b/d in 1980. Even this amount might disappear if
        China's domestic demand grows more rapidly than future gains
        in her oil production.

C394   "China's Energy Performance," by Vaclav Smil. Current History
        429 (Sept. 1977):63-67. tables.
            Chronicles major events in 1976 and 1977 within China's
        energy industries. Gains of 13% and 11% were registered in
        the production of crude oil and natural gas, and some tech-
        nological innovations were achieved in drilling equipment and
        refining facilities. Smil also reports that there is major
        expansion in China's petroleum processing industries: large-
        scale construction for petrochemical complexes is underway in

Canton, Shanghai, and Shenyang, and new facilities are being
added in the Peking and Shengli complexes and the Tach'ing
general petrochemical works. The author also describes key
improvements in China's oil port facilities and the building
of larger tankers in Chinese shipyards.

C395   "Getting Oil at High Speed--Notes on Taching (1)." China
        Reconstructs 26, no. 9 (Sept. 1977):14-20. illus. (some
        color).
            Contains a colorful sketch and moving stories about the
        Tach'ing oil field. It describes the oil workers' zeal to
        "grasp revolution and promote production," and evaluates ten
        major achievements made in the course of Tach'ing's growth.
        It also reports on Drilling Team No. 1205, which broke world
        records for drilling footage in a year, and briefly mentions
        the emphasis on research and development that characterizes
        the Tach'ing oil complex.

C396   "The Story of Iron Man Wang." China Reconstructs 26, no. 9
        (Sept. 1977):24-27. illus.
            A brief biographical sketch of a hero, Wang Chin-hsi, dur-
        ing the development of Tach'ing. Wang's leadership and driving
        spirit is credited with increasing the speed with which oil
        wells have been drilled. The article tries to prove the Maoist
        theory that human factors are more important than machines, an
        ideology that has superseded other considerations in the build-
        ing of China's oil base.

C397   "Worker-Peasant Villages--Notes on Taching (2)." China
        Reconstructs 26, no. 9 (Sept. 1977):21-23. illus.
            This second article on the Tach'ing oil field describes an
        on-site visit to the sprawling agricultural settlements around
        the oil field and its petrochemical complex. It also gives a
        sketch of daily life life for the 250,000 residents of the
        worker-peasant villages. The settlements are complete with
        farms, hospitals, schools, small factories, stores, and admin-
        istrative offices.

C398   "Criticism of Teng Hsiao-ping Is in a New Upsurge in Taching
        Oil Field." Chinese Economic Studies 9, no. 1 (Fall 1977):
        16-21.
            A description of the mass movement throughout the
        Tach'ing oil field to criticize Vice-Premier Teng for
        allegedly being a capitalist roader. More than 32,000 workers
        in the area assembled to accuse Teng of following a revision-
        ist line. Meanwhile, the masses were also exhorted to "grasp
        revolution and promote production." As a result, production
        of oil has been boosted, potential resources tapped, and the
        highest production record set ever.

C399   "Learning from Ta-ch'ing: China's Oil Prospects," by
        W. Klatt. Pacific Affairs 50, no. 3 (Fall 1977):445-59.
            This article has two parts. In the first, Klatt sum-
        marily reviews five books written on Chinese oil, recommending
        Chu-yuan Cheng's China's Petroleum Industry (entry B32) and
        Vaclav Smil's China's Energy (entry B29) as the most compre-
        hensive and precise. The second part deals with the prospects
        and potentials of China's crude oil production, including his
        own assessment of output from China's leading oil fields.

C400   "China Buys Two Drilling Rigs." Petroleum Times 81, no. 2059
        (Sept. 16, 1977):6.
            Reports on China's acquisition of two offshore rigs of a
        semisubmersible type from Japan and Norway. The Norwegian
        rig, named Borgny Dolphin, is capable of drilling in up to 600
        feet of water, while the Japanese-built rig is of ETA design
        and can explore waters up to 300 feet deep. The capability
        of these rigs is said to be substantially higher than the ones
        currently employed in Chinese waters.

C401   "Japan Sends Oil Mission to China." Petroleum Times 81,
        no. 2059 (Sept. 16, 1977):6.
            Reports that the representative of a major oil firm,
        Idemitsu Kōsan, has informed the Chinese that if the quality
        and quantity of China's oil are similar to that of the Middle
        East, Japan is ready to purchase 100 million tons of crude oil
        from China. Mr. Tominaga indicates Japan's intention to di-
        versify the sources of her oil imports.

C402   "U.S. Urged to Oppose Oil Claim by Taiwan," by Drew Middleton.
        New York Times (Sept. 18, 1977):13.
            Reports on a study released by the Carnegie Endowment for
        International Peace (entry B39). The study, prepared by
        Selig S. Harrison, urges Washington to disassociate itself
        from Taipei on offshore exploration off the coast of Taiwan
        and China. The author recommends the U.S. recognize Peking's
        claim to her continental shelf, which extends 50 miles off the
        coast of the mainland.

C403   "Chinese Energy Demand Falls Back," by Vaclav Smil. Energy
        International 14, no. 10 (Oct. 1977):27-29. map.
            Reviews China's overall energy performance in 1976.
        Despite serious earthquake damage to energy centers, the oil
        and gas sectors performed creditably, scoring a 13% gain in
        crude oil output and an 11% increase in natural gas produc-
        tion. Numerous technological innovations have been accom-
        plished in the oil and gas extracting and refining industries.
        China's major fields, including Tach'ing and Shengli, recorded
        steady production increases. The expansion of existing and
        the building of new processing facilities continued. Major
        improvements have also been made on port facilities to handle
        increased oil traffic.

C404   "Energy Solution in China," by Vaclav Smil. <u>Environment</u> 19,
no. 7 (Oct. 1977):27-31. illus.
China has developed thousands of small-scale bio-gas
facilities to produce the alternate fuel to replace oil and
gas for power generation.

C405   "Awaiting the Rush of Chinese Crude, If Any--China '77," by
George Lauriat. <u>Far Eastern Economic Review</u> 98, no. 40
(Oct. 7, 1977):67-69. illus.
Summarizes news reports about the development of the Chinese petroleum industry in 1977. Analyzes and evaluates oil
production, the construction of new fields, progress in offshore drilling, procurement of offshore rigs and their in-
tended deployment, refining capacities, and the gap between
oil output and refining capability. Also focuses on China's
oil export markets and the prospects for increases in her
output of natural gas.

C406   "Petroleum Industry Keeps Advancing." <u>Peking Review</u> 20,
no. 41 (Oct. 7, 1977):44.
Crude oil production in the first 8 months of 1977 shows
a 10% increase and natural gas a 24% gain compared with the
corresponding period last year. A group of new oil and gas
fields are being opened up. The oil equipment manufacturing
industry has likewise attained its production targets.

C407   "China Ties Industrial Future to Taching Oil Plant," by
Harrison E. Salisbury. <u>New York Times</u> (Oct. 14, 1977):All.
An American journalist's impressions of Tach'ing as an
industrial complex combining oil refineries, agricultural
enterprises, fertilizer plants, and such auxiliary services
as schools, shops, and hospitals. He is impressed by the
integration of the industrial establishments with agribusiness
enterprises and the disciplined life, which is held up as an
institutional model to Chinese all over the country.

C408   "New China's Natural Gas Industry." <u>Economic Reporter</u>
(English supp.) 4 (Oct./Dec. 1977):43.
A brief review of the development of China's natural gas
industry in the past 30 years. The article claims that since
the beginning of the Cultural Revolution the output capacity
of China's natural gas has gone up by 5.7 times and the rate
of opening up gas fields has increased by 200%.

C409   "Taching Chemical Fertilizer Plant Goes into Operation."
<u>Economic Reporter</u> (English supp.) 4 (Oct./Dec. 1977):16.
illus.
A giant chemical fertilizer plant went on line at the
Tach'ing oil base in 1976. Using natural gas produced at
Tach'ing as feedstocks, the plant has the capacity of one
million tons of standard chemical fertilizer a year.

C410   "How Rolligon Sold Petroleum Equipment to China." U.S.-China Business Review 4, no. 6 (Nov./Dec. 1977):3-4. illus.
　　　The Rolligon Corp. of Houston, Texas has signed a contract with China for the sale of several million dollars worth of rough-terrain vehicles and spare parts designed to support oil drilling operations. American businessmen must be patient and persistent in their negotiations with Chinese representatives if they are to succeed in penetrating the Chinese market.

C411   "China Reports Expansion of Shengli Oilfields." New York Times (Dec. 4, 1977):8.
　　　Hsinhua press agency reports that the Shengli oil field has been undergoing major expansion and is fast becoming an integrated oil center complete with prospecting, exploration, petroleum production, transportation, oil research, and petroleum engineering.

C412   "China to Buy Two U.S. Platform Rigs from Houston Firm." Oil and Gas Journal 75, no. 51 (Dec. 12, 1977):35.
　　　Reports on China's purchase of two offshore platform rigs, valued at $15 million, from the Houston-based National Supply Co. These rigs will likely be used to expand undersea exploration of her vast oil deposits on the continental shelf.

C413   "China's Energetics: A System Analysis," by Vaclav Smil. In Chinese Economy Post-Mao, by the Joint Economic Committee, U.S. Congress, vol. 1, Policy and Performance. Washington, D.C.: Government Printing Office, 1978, pp. 323-69. illus., maps, tables.
　　　Presents a comprehensive assessment of China's energy resources, demand, and supply, including oil and gas. Also forecasts China's energy potentials in the 1980s and 1990s.

C414   "China's Mineral Economy," by K.P. Wang. In Chinese Economy Post-Mao, by the Joint Economic Committee, U.S. Congress, vol. 1, Washington, D.C.: Government Printing Office, 1978, pp. 370-402. illus., maps.
　　　Surveys overall mineral development in China, with a brief profile of the current state of oil and gas production.

C415   "On the Feature of Turbidite Sequences in Some Regions of China," by L. Jiliang et al. Scientia Geologica Sinica 1 (1978):26-44. Abstracted by Geo Abstracts E: Sedimentology, no. 6 (1978):510.
　　　Analyzes the sedimentary geology of northwest China from Mt. Ch'inling in the west to Chekiang province on the east coast.

C416   "China's Big Potential for Offshore Oil," by A. Rahmer. Petroleum Economist 45, no. 1 (Jan. 1978):9-10.
　　　Cites Selig Harrison's book on China's offshore oil potential (entry B39) and the risk of China's territorial disputes with her neighbors.

C417   "Oil and Gas in China (I)," by Robert W. Scott. <u>World Oil</u> 186 (Jan. 1978):85-88. color illus., maps.

     The first of two reports (see also entry C459) on the Shengli oil field by the editor of <u>World Oil</u>. The author recently spent seventeen days leading a trade delegation to this oil field. He gives a brief sketch on its location, unique features, operations, and workforce. Fourteen full-color photos provide a vivid and rare glimpse of China's second largest oil field in action.

C418   "An Up and Coming Oil Refinery." <u>China Reconstructs</u> 27, no. 1 (Jan. 1978):27-29. illus.

     A sketch of the Changlin Oil Refinery in Hunan province, which was built in 1971. For the past 7 years, the output and quality of its petroleum products have been improved steadily. The capacity of the plant has been raised 20%, and the profits have tripled the amount of its original investment.

C419   "Trading U.S. Technology for Chinese Oil." <u>Business Week</u>, no. 2518 (Jan. 23, 1978):48.

     Comments on the possibility of extending the U.S. aid to China in the form of oil technology and credit, making China a major factor in the world oil market.

C420   "Chinese VCM Unit to Use Hoechst-Goodrich Route." <u>Chemical Marketing Reporter</u> (Jan. 30, 1978):5.

     Reports that a vinyl chloride plant employing the Hoechst-Goodrich process has gone on line in China. This is the first industrial plant in China to produce VCM from ethylene. Uhde, a West German concern, has been responsible for engineering, purchase of equipment, and supervision of construction. The plant's rated capacity is 80,000 tons a year.

C421   "China's Offshore Petroleum," by Jan-Olaf Willums. <u>China Business Review</u> 5, no. 1 (Jan./Feb. 1978):31-35. illus., maps.

     An evaluation and prediction about China's offshore oil by an American-trained Norwegian geologist. Willums reviews the hydrocarbon potentials in China's continental shelf through analyses of her offshore oil deposits, compares the potentials with constraints on their development, and discusses the possibility of technological transfers from the West in light of China's past and present oil policies. If China relies on her own knowhow, he predicts, no more than 1 million b/d of oil can be extracted from her offshore deposits by 1985. If China imports Western technology, however, total offshore and on-shore production could reach 330 million tons by 1985. The market for Western oil equipment in 1980, Willums concludes, will be worth $1 billion.

C422　"The Shanghai No. 1 Petroleum Machinery Plant." <u>China Business Review</u> 5, no. 1 (Jan./Feb. 1978):34.
　　　　An account of an inspection tour by a delegation from the U.S. National Council for U.S.-China Trade. The members observe the Shanghai No. 1 plant, which builds drilling machinery for the petroleum industry and mining equipment for mineral extraction. The facility is of medium scale and employs 1,700.

C423　"Shengli Journal," by Stephanie Green. <u>China Business Review</u> 5, no. 1 (Jan./Feb. 1978):31-35. illus.
　　　　An account of a rare visit to the Shengli oil field, the second largest oil base in Shantung province, by a delegation from the U.S. National Council for U.S.-China Trade. Observations about operations of the oil field are made by different members of the visiting group, which includes many American oil equipment manufacturers' representatives. Generally, Chinese operations are said to be 20-30 years behind those of the U.S. Members of the group estimated that oil output is around 200,000 b/d.

C424　"Tachai and Taching," by Dipankar Gupta and Harmala Kaur Gupta. <u>China Report</u> 1 (Jan./Feb. 1978):3-14.
　　　　Analyzes the significance of Tachai and Tach'ing as models for China's agricultural and industrial development. The authors maintain that through the application of the dialectical unity between revolution and production, the use of political persuasion and sheer human spirit as driving forces, China has succeeded in building an independent oil industry, as exemplified by Tach'ing.

C425　"China Sets New Sights," by Leonard LeBlanc. <u>Offshore</u> 38, no. 2 (Feb. 1978):91-93. illus.
　　　　Chinese oil officials have set a target for oil production of 10 million b/d before the end of the century, a 600% increase over the current rate. A large portion of projected increase will come from offshore explorations.

C426　"Danger Zones in the South China Sea," by Selig Harrison. <u>Petroleum News--Southeast Asia</u> (Hong Kong) 8, no. 11 (Feb. 1978):26-30.
　　　　Comments on his book <u>China Oil and Asia: Conflict Ahead?</u> (entry B39). The attention is focused on variations between his prediction of China's crude oil export potential and the CIA's estimate. Also, potential territorial conflicts with her neighbors over offshore claims are discussed.

C427   "East Asia and Australia: A Remarkable Development, China,"
        by A.A. Meyerhoff and J.O. Willums. Norwegian Oil and Gas
        Journal (text in English and Norwegian) 2 (Feb. 1978):43-45.
            The authors provide a general survey of the development of
        China's petroleum industry over the past twenty years and then
        appraise both the onshore and offshore hydrocarbon potentials
        of China through geological analysis.

C428   "Much Slower Growth." Petroleum Economist 45, no. 2 (Feb.
        1978):74-75. map.
            Analyzes the sluggish performance of China's oil industry
        in 1977. A sharp decline to 13% from previous annual gains of
        more than 20% in production, prompts the Economist to ponder
        on the possible causes for the slowdown. Its only explana-
        tions are that the production of major existing oil fields has
        peaked and that it takes time to develop entirely new oil re-
        sources such as the offshore fields.

C429   "The Shengli Oil Field." China Reconstructs 27, no. 2 (Feb.
        1978):42-46. color illus.
            A profile of the second largest oil base in China.
        Located in northern Shantung province, the oil field was dis-
        covered in 1965. Despite the complicated nature of geological
        strata, development has proceeded rapidly to make it the sec-
        ond largest oil field (after Tach'ing) in China. An account
        is given of a heroic drilling team, No. 3252, which set a
        world record by sinking 20 high-yielding wells in a year.

C430   "Oil for the Lamps of Tokyo?" by James Cook. Forbes 121,
        no. 3 (Feb. 6, 1978):37-38. map.
            A general discussion on the production and export poten-
        tials of China's oil, particularly sales potentials to Japan.
        The author maintains that oil has become the best way for
        China to earn the substantial foreign exchange she needs to
        purchase Western technology for industrialization. The cur-
        rent production volume is not sufficient to provide much for-
        eign currency. In order to exploit her oil resources more
        rapidly, the author concludes, China has to liberalize her
        rigid policy of "self-reliance" and allow foreign equity in
        the development of her offshore deposits.

C431   "Potential Giant: Peking Experts Visit U.S." Time 111, no. 6
        (Feb. 6, 1978):57. illus., map.
            The visit to the U.S. of sixteen Chinese oil experts is an
        indication that China wants to use Western technology to
        rapidly develop its oil potentials. Vast parts of China, both
        onshore and offshore, have not yet been geologically surveyed,
        and there is a possibility that China can become an oil giant
        if Western knowhow is applied to explore and develop its oil
        deposits. China aims at a fourfold increase in crude oil pro-
        duction, to 8 million b/d, by 1990.

C432   "Natural Gas Chemical Plant Built." Peking Review 21, no. 8
       (Feb. 24, 1978):30.
           A modern chemical plant using natural gas feedstocks has
       been built and put on line in Yunnan province. The plant's
       rated capacity is 300,000 tons of synthetic ammonia and
       480,000 tons of urea per year--equivalent to 1 million tons
       of standard chemical fertilizer. Construction of the plant
       started in October 1974, and by September 1977 production
       trials commenced. The plant is one of the most highly auto-
       mated in China.

C433   "Second Plant Finished for Taiwan Firm." Oil and Gas Journal
       76, no. 9 (Feb. 27, 1978):63. illus.
           Reports that an acrylonitrile plant has been built for the
       China Petrochemical Development Co. in Taiwan by Badger, Ltd.
       Badger had built a similar plant in Taiwan in 1975.

C434   "China: Toward a Realistic Assessment," by J.O. Willums and
       A.A. Meyerhoff. Petroleum News--Southeast Asia (Hong Kong) 8,
       no. 12 (Mar. 1978):23-25.
           Unless more seismic data and geological surveys are made
       available, China will be known only as a nation with vast oil
       potentials. The authors estimate that a production level of
       6-7 million b/d by 1985 is a realistic goal that China could
       attain.

C435   "China's Energy Resources," by Chu-yuan Cheng. Current
       History 435 (Mar. 1978):121-24, 136-38. tables.
           The author's second evaluation of China's energy resources
       to appear in Current History. He describes the rapid expan-
       sion of oil and gas production and the rise in their share of
       overall Chinese energy production from 3% in 1957 to 33% in
       1976. Cheng asserts that China's political instability in
       1976 diminished the growth rate of crude oil output in that
       year from an annual average of 20% to a mere 13%. The author
       also expresses some doubts concerning the validity of output
       figures released by Chinese government media.

C436   "George Bush: Oil for China's Self-Reliance," by Willard C.
       Rappleye, Jr. Financier 2, no. 3 (Mar. 1978):28-30.
           An interview with George Bush, former chief of the U.S.
       Liaison Office and former CIA head, following his return from
       a visit to China. Bush maintains that oil is the key to
       China's economic development and that the present leadership
       is pragmatic enough to do business with U.S. oil companies in
       order to develop the country's oil resources. He also asserts
       that as China will not allow foreign ownership of her natural
       resources, some forms of compromise will have to be worked out
       with American companies to provide them with incentives for
       their capital investments and services. This could mean some
       sharing of China's oil output.

C437   "Japan, Final Terms for Chinese Oil." Petroleum Economist 45,
        no. 3 (Mar. 1978):123.
            Reports on Sino-Japanese accord on 1978-1985 importation
        of Chinese crude.

C438   "Outline of the Tectonic Evolution of Southwestern China," by
        P.F. Fan. Tectonophysics 4 (Mar. 3, 1978):261-67.
            An analysis of five tectonic units in southwestern China:
        (1) the east K'unlun fold belt, (2) the Yangtze paraplatform,
        (3) the south China paraplatform, (4) the western Yunnan fold
        zone, and (5) the Himalayan fold zone.

C439   "Asian Alliances and Chinese Oil," by Jeremiah Novak.
        America 8 (Mar. 4, 1978):165-67. illus.
            Presents a bold and novel hypothesis about America's
        changing role in the geopolitics of northeast Asia in view of
        the Chinese discovery of vast oil resources. Novak asserts
        that the economic renaissance of Japan, South Korea, and
        Taiwan is providing the U.S. with a chance to realign and
        redirect the superior productive resources of these three
        countries toward the vast Chinese market. This would divert
        their finished products from the U.S., which is already suf-
        fering large trade deficits.

C440   "China Probably Won't Be OPEC Buster," by John Freeman. Japan
        Times (International ed.) (Mar. 4, 1978):12.
            Summarizes the consensus of oil experts attending a con-
        ference in Singapore, who believe China will export just
        enough crude oil to meet her foreign exchange needs. China
        will be a minor factor in the world oil trade before 1985,
        since her production capability cannot be expected to increase
        substantially in the near future.

C441   "Shengli Crude is Key Component in Warm Peking-Manila Rela-
        tions." Business China 5 (Mar. 8, 1978):30-31.
            Discusses the oil connections between China and the
        Philippines, which has become the second major importer of
        Chinese crude, largely because of low prices charged for oil
        produced at the Shengli oil field. In 1978 China is expected
        to export one million tons of oil to the Philippines, as com-
        pared with only half that much in 1975.

C442   "China Fuels Oil Industry Hopes," by Ranjit Gill. Far
        Eastern Economic Review 99, no. 10 (Mar. 10, 1978):82.
            As China starts to accelerate her development of both
        onshore and offshore oil resources, American oil field equip-
        ment manufacturers place their hopes on her future purchases.
        China is said to be in need of specialized pipeline equipment,
        sophisticated drilling and production testing equipment, and
        oil well safety apparatuses. She is capable of manufacturing

70% of her needed oil equipment, but has a good appetite for advanced foreign technology.

C443   "The Geopolitics of China's Oil," by Henry Jackson. New York Times (Mar. 25, 1978):19.
        Jackson, chairman of the Senate Committee on Energy and Natural Resources, asserts that the development and expansion of China's oil production is not only assisting the U.S. in maintaining a balance of power in the world, but is also in the U.S. national interest. He urges the U.S. government to take an active part in influencing China's formulation of her offshore policy, and considers the opportunity of assisting in the development of her undersea oil an important chance to improve U.S.-China relations.

C444   "New Technique in Oil Drilling Popularized." Peking Review 21, no. 13 (Mar. 31, 1978):30-31.
        The Ministry of Petroleum is conducting classes at North China Oil Field to popularize a new drilling technique--high pressure jet oil drilling. It was developed in the 1960s and perfected in the 1970s. All key personnel from other oil fields across the country are assembled here to attend the lectures on this new technology.

C445   "Feasible New Petrochemical Projects in the ROC," by Jerome S.N. Hu. Economic Review (International Commercial Bank of China, Taipei), no. 182 (Mar./Apr. 1978):1-4. tables.
        Hu, chairman of the Chinese Petroleum Corp., outlines general trends of the petrochemical industry in Taiwan and possible manufacturing of new products such as acetone, phenol, vinyl acetate, and acrylate. He made his comments in a speech at the joint meeting of R.O.C.-U.S.A. Economic Council held in Chicago on October 18-19, 1977.

C446   "China's Oil Extraction Techniques May Soon Be Exported to Japan." Business China 7 (Apr. 5, 1978):48.
        Discusses the possibility that Chinese technical knowhow and innovations in oil exploration and production may be exported to Japan.

C447   "U.S. Catches China's Eye," by Melinda Liu and Peter Weintraub. Far Eastern Economic Review 100, no. 17 (Apr. 28, 1978):37-38. illus.
        For the first time, a U.S. firm has succeeded in selling a new offshore rig to China. Bethlehem Singapore, an affiliated firm of Bethlehem Steel, has signed a contract estimated at $20-30 million for the sale of a mat-type jackup unit, capable of operating in waters up to 250 feet deep. Presumably, this new rig will be used in the waters near Hainan island off Kwangtung province in the South China Sea. The delivery of the rig will be in early 1979.

C448   "Fair Prospects." <u>Petroleum Economist</u> 45, no. 5 (May 1978):
        217-18.
            A summary of the assessment of China's oil deposits by
        A.A. Meyerhoff and Jan-Olaf Willums, published by Petro-
        consultants of Geneva. According to the report, China's
        recoverable oil reserves are estimated to be about 9.5 billion
        tons, of which 4.1 billion are located offshore. Ultimate
        recoverable gas is estimated to be about 200 trillion cubic
        feet, half of it from offshore. The authors consider it
        reasonable for Chinese production to reach 6-7 million b/d or
        320 million tons of oil by 1985 (about half from offshore) if
        China's output can continue to grow at a 16% annual rate dur-
        ing the next eight years. (The original assessment is avail-
        able from the authors for $500.)

C449   "An Old Szechwan Asset for New Markets," by F. Armentrout.
        <u>Petroleum News--Southeast Asia</u> (Hong Kong) 9, no. 2 (May
        1978):20-22.
            Stresses the magnitude of the Szechuan basin in the pro-
        duction of natural gas in China. Also provides a general
        survey of the Chinese natural gas industry.

C450   "Solving the Chinese Oil Puzzle," by G. Segal. <u>Petroleum</u>
        <u>Review</u> 32, no. 377 (May 1978):53-56.
            The petroleum industry's managerial structure has been
        held up as a model for all other industrial organizations in
        China to emulate. The political slogan "Learn from Tach'ing"
        is widely studied.

C451   "Taiwan and Korea Load Up for Exports." <u>Chemical Week</u> 122
        (May 3, 1978):28-34.
            Reports on Taiwan's petrochemical industry and its success
        in exporting petrochemical products.

C452   "U.S. Suppliers Strengthen Their Share of China Oil Rig
        Market." <u>Business China</u> 9 (May 3, 1978):61.
            Analyzes the prospects of U.S.-made offshore oil rigs to
        China in the first 5 months of 1978. Opportunities seem to
        have brightened for the U.S. manufacturers.

C453   "China Discovers Offshore Oil Field." <u>Washington Star</u>
        (May 6, 1978):A4.
            Kwangtung provincial radio reports the discovery of oil
        and natural gas in the South China Sea, as well as plans to
        develop the area into a major oil field.

C454   "New Oil Discoveries Reported off China: Major Field
        Planned," by Fox Butterfield. <u>New York Times</u> (May 6, 1978):
        29.
            Reports from Hong Kong on monitoring China's announcements
        on the discovery of offshore oil in the South China Sea. No

details have been disclosed by the Chinese on the dimensions of the find. But the reporter surmises the location to be either off the Gulf of Tonkin and the southeast coast of Hainan or at the mouth of the Pearl River in Kwantung province. The *Times* correspondent further reports that the Chinese plan to develop the new offshore finds into a major oil field.

C455   "Marathon to Build Two Jack Ups for Mainland China." Oil and Gas Journal 76, no. 21 (May 22, 1978):43.
        China has placed an order for two shallow jackup rigs from Marathon Manufacturing Co. of the U.S. They will be constructed at the company's Singapore shipyard and will be capable of working in waters up to 250 feet deep, withstanding winds up to 80 miles per hour, and wave height up to 30 feet.

C456   "Black Gold." VISTA (Taiwan) 3 (May/June 1978):31-33. illus.
        Describes how a wildcat well drilled by the Chinese Petroleum Corp. of Taiwan struck oil and gas on April 25, 1978 along the west coast of central Taiwan. This is the first oil well on Taiwan in many years to produce significant quantities of oil and gas (about 680 b/d of crude oil).

C457   "Taching--China's Model for Industry," by Lois W. Snow. Journal of World Trade Law 12 (May/June 1978):228-40.
        A profile of the Tach'ing oil field and its petrochemical complex by the widow of the late Edgar Snow.

C258   "CPC (Chinese Petroleum Corporation) Goes Petrochemical," by Wilfred Brown. Free China Review 28 (June 1978):20-23. illus.
        Describes Taiwan's all-out effort to upgrade her petroleum industry by investing huge amounts of capital in downstream operations.

C459   "Oil and Gas in China (II)," by Robert W. Scott. World Oil 186 (June 1978):101-11. color illus.
        The second of two reports (see also entry C417) on a visit to the Shengli oil field. It provides more detailed information from on-the-spot talks with the operating personnel and insightful observations on the oil field operations. The report identifies the producing areas of the oil fields; discusses Chinese philosophy on the oil field management; discerns managerial attitutde; and describes drilling, rigs, and oil production. Finally, the author lauds Chinese accomplishments on the oil and gas front and discerns that China is well aware of the need for acquiring advanced technology to speed up her oil production.

C460   "The Oil Extraction Research Institute of Taching Oilfield,"
       by Chang-lu Cheng. China Pictorial 6 (June 1978):6-8. illus.
           A portrayal of the growth and development of a petroleum
       research institute at Tach'ing. Since its inception in 1962,
       this once small institute has now grown into a substantial
       establishment with 5 departments and a staff of 425. The
       institute so far has contributed nearly two thousand important
       items, some of which are advanced by world standards, such as
       the hydraulic packer.

C461   "Striving for Ten New Tachings." China Trade Report 16
       (June 1978):2-3. illus.
           Analyzes and assesses the strategies behind Peking's call
       for the building of ten or more Tach'ing-type oil fields. The
       Report contends that beyond China's effort to earn foreign
       exchange for her oil, Peking aims her "petro-diplomacy" at
       oil-hungry neighbors to gain friends and improve relations
       with countries in Asia. China's emphasis on the development
       of her offshore oil has led to the speculation that in the
       not-too-distant future China may be more sympathetic to
       production-sharing, which is common between the oil producing
       countries and foreign oil concerns.

C462   "Man-Made Diamond Drill Bits." Peking Review 21, no. 23
       (June 9, 1978):31.
           The production of artificial diamond drill bits has been
       developed and perfected by a research group headed by Lin
       Tseng-tung of Peking Research Institute of Powder Metallurgy.
       The man-made bits will help improve techniques of geological
       prospecting in China, which set a record by boring through
       350 meters of hard stratum in 1976.

C463   "Offshore Reserves Top PRC Priority." Offshore 38, no. 7
       (June 20, 1978):188, 190. map.
           Exploration and exploitation of oil in China's continental
       shelf is the first priority in China's energy policy. The
       U.S. analysts estimate China's offshore oil potential equaling
       that of onshore reserves.

C464   "Taiwan's Gas Hopes Evaporate," by William Kazer. Far Eastern
       Economic Review 100, no. 25 (June 23, 1978):100. illus.
           With the demand for natural gas increasing at 15% a year,
       far outstripping the expansion in the production of natural
       gas in Taiwan, the major gas supplier, Chinese Petroleum Corp.,
       is asking manufacturers who are heavy gas consumers to convert
       to the use of heavy oil by the end of 1978. Proved natural
       gas reserves in Taiwan will last only 11 years at the current
       rate of consumption, and additional supply is costly and hard
       to come by.

C465 "Japan-China Trade Up 40% with Plastic Resins Leading the
      Way." Chemical Marketing Reporter (July 3, 1978):4, 20.
           The shipment of Japanese chemicals to China, led by
      plastic resins, rose 40% during the first quarter of 1978 from
      its 1977 level. A variety of chemical products are responsi-
      ble for the surge in Japanese exports to China. They include
      phthalic anhydride ($3.7 million), benzene ($4.3 million),
      and plastic resins ($15 million), up 64% from last year's
      record.

C466 "German's Uhde Seen Winner of Key Chinese Contracts."
      European Chemical News (London) 32, no. 844 (July 7, 1978):4.
           West Germany's Uhde is the apparent winner of five major
      petrochemical contracts involving the building of a 60,000
      ton per year polyethylene plant and a 200,000 ton per year
      ethanol plant to be constructed in the Tach'ing petrochemical
      complex.

C467 "JGC/Marubeni Win Chinese Ethylene Plant Contract." European
      Chemical News (London) 32 (July 14, 1978):35.
           Reports that Japan's major trading firm, Marubeni, has
      obtained a contract to build a chemical plant in China.

C468 "Oil Broadens Chinese Development Role," by A.A. Meyerhoff and
      J.O. Willums. Oil and Gas Journal 76, no. 29 (July 17, 1978):
      91-98. illus., tables.
           Probably one of the latest and most authoritative geologi-
      cal analyses of China's onshore and offshore oil reserves and
      exploration by two renowned geologists. Listing the major
      oil-bearing basins in China, the authors attempt to estimate
      ultimate oil recoveries from China, by basin, and also the
      hydrocarbon content of China's outer continental shelf. The
      authors further make a geological evaluation of China's key
      oil fields and estimate their current rate of production.
      And finally, they weigh hydrocarbon potential in the onshore
      and offshore regions of China and discuss briefly China's oil
      transportation by pipeline and tanker.

C469 "Peking Informally Seeks Assistance of Japanese Business in
      Oil Probes," by Takahiro Okada. Japan Economic Journal
      (July 18, 1978):5.
           For the first time in recent memory, China has approached
      Japanese business leaders about joint development of China's
      oil resources. The proposal for assistance was made to a
      mission of Japan Committee for Economic Development (Keizai
      Doyukai) when it was visiting China in May 1978.

C470   "4 U.S. Oil Companies, with Official Support, Start Talks
        with China," by Richard Burt. New York Times (July 20, 1978):
        A4+.
            With Pennzoil Co. of Houston taking the lead, four U.S.
        oil companies (also including Exxon, Union Oil of California,
        and Phillips Petroleum) have entered into negotiations with
        Chinese authorities on the possibility of U.S. participation
        in China's offshore exploration.

C471   "Scientific Research at Taching." Peking Review 21, no. 29
        (July 21, 1978):30. illus.
            Scientific research endeavors at the Tach'ing oil base
        have resulted in 32,300 technical innovations and research
        achievements since 1960. Among these are a new method for
        separating p-xylene from xylene and a technique for maintain-
        ing stable oil flow from wells. Currently, Tach'ing has 2
        design institutes and 23 research institutes with a staff of
        6,000.

C472   "U.S. Oil Firms May Get Chance to Help China Exploit Its Off-
        shore Deposits," by Barry Kramer. Wall Street Journal
        (July 21, 1978):28.
            The Journal's Hong Kong correspondent reports that with
        open encouragement from the U.S. government, major American
        oil companies are poised to assist China in developing her
        vast offshore oil deposits. China has issued an invitation
        to four U.S. oil companies, Exxon, Pennzoil, Phillips Petro-
        leum, and Union Oil of California, to send delegations to
        Peking for discussions on offshore development. There are
        indications that China might be willing to offer contractual
        share of oil produced from the undersea oil wells.

C473   "Japan, China Plan Offshore Venture," by A.E. Cullison.
        Journal of Commerce 337, no. 2425 (July 25, 1978):1, 32.
            According to a Japanese oil mission visiting Peking,
        China and Japan have tentatively agreed on joint development
        of offshore oil reserves in the Pohai Gulf, in waters off the
        coast of eastern China, and at the mouth of the Pearl River
        near Hong Kong. If a formal contract is signed, this will be
        the first time China has allowed foreign firms to participate
        in the development of her offshore oil resources. Although
        the Japanese success in negotiations with the Chinese does
        not preclude U.S. participation at a later time, the Japanese
        now seem to enjoy an edge.

C474   "Tokyo, Peking to Explore Oil Offshore China." Wall Street
        Journal (July 25, 1978):5.
            A basic agreement has been reached between China and the
        government-owned Japan National Oil Corp. for the joint devel-
        opment of undersea oil deposits in the Pohai Gulf. Both

countries are also studying the feasibility of joint offshore
exploration in the South China Sea near the mouth of the
Pearl River in Kwangtung province.

C475   "China Reveals Long-Term Energy Development Plans," by Vaclav
        Smil. Energy International 15, no. 8 (Aug. 1978):23-25, 29.
        illus., map.
            An analysis and appraisal of the long-rance energy plans
        revealed by China's new leadership. Coal will remain the
        mainstay of China's energy sources, with oil gradually expand-
        ing its share of the energy supply. China's new leadership
        calls for the discovery and construction of ten or more
        Tach'ing-type oil and gas fields. Rapidly rising consumer
        demands--particularly for the mechanization of agriculture
        and the modernization of transportation--will require China to
        upgrade and modernize her petroleum industry.

C476   "Japan and China Will Take Up Development of Oil." Japan
        Economic Journal 16, no. 813 (Aug. 8, 1978):5, 19.
            China's Deputy Premier Kang Shien and a high-powered
        Japanese technology mission have reportedly reached a tenta-
        tive agreement for the joint development of oil resources off
        the coast of China in the Pohai Gulf and along the Zam Jiang
        River (Pearl River?) in southern China. This is the first
        time that China is said to have agreed on direct foreign par-
        ticipation in the development of her oil resources.

C477   "China's Oil Fields Hold Promise of Foreign Participation."
        Business China 15 (Aug. 9, 1978):107-8.
            Discusses the "meaning" of foreign participation in
        China's oil search. Suggests that the Chinese might be ready
        for some forms of production sharing--with foreigners supply-
        ing the financing, technology, and machinery in return for a
        portion of the oil produced by joint ventures.

C478   "China--U.S. Oil Talks; Peking in All-Out Hunt for Best Tech-
        nology," by Hobart Pouen. Washington Post 249 (Aug. 11,
        1978):1-A8.
            China's Liaison Office in Washington has contacted State
        Department officials to invite four American oil companies,
        Exxon, Phillips Petroleum, Pennzoil, and Union Oil of Cali-
        fornia, to send delegations to Peking to negotiate the devel-
        opment of China's offshore oil deposits. Negotiations are
        likely to lead to an investment of $2-$50 billion by American
        companies in the shared development of Chinese oil.

C479   "Split with China Jolts Albanian Oil Industry." Oil and Gas
        Journal 76, no. 33 (Aug. 14, 1978):168.
            The termination of Chinese economic aid, the withdrawal of
        Chinese oil experts, and the cessation of training for

Albanian petroleum technicians have dealt a blow to the developing oil industry of Albania, as illustrated by a long delay for the opening of the country's largest oil refinery--Ballsh integrated works. The production target for oil in Albania's current five-year plan has not been met, and her crude oil output in the last two years has been static at best.

C480   "Oiling the Doors." Economist 268 (Aug. 19, 1978):61. illus.
        China's huge appetite for capital goods and technology draws the Economist's attention to China's steel and petrochemical projects. Since 1978 China is said to have been building more than 30 petrochemical and fertilizer plants using Western technology and still more are being planned. Contracts signed with foreign firms include a 300,000-ton-per-year ethylene plant at the Tach'ing oil field.

C481   "Esso: Far East Oil Potential Limited." Oil and Gas Journal 76, no. 35 (Aug. 28, 1978):37-38. graph.
        An account of the summary of the Circum-Pacific Energy and Mineral Resources Conference held in Honolulu. Among the focal points of discussion at the meeting were the petroleum outlook in East Asia, exploration activity in the area, geothermal energy, and the prospect of oil in China. Many geologists, including A.A. Meyerhoff of Tulsa, Oklahoma, and Otis O. Fox of Esso Exploration, Inc., agreed that it is questionable whether China can fully develop her oil potentials, meet all internal needs, and still produce a substantial surplus that would affect the global oil supply.

C482   "China Offering Its Oil for Sale in the U.S. Market," by J.P. Smith. Washington Post (Aug. 30, 1978):1, A13.
        Gulf Oil is reported to have been approached by an intermediary to consider the purchase of Chinese crude. Reports also are circulating about a barter deal with an American automobile firm for the exchange of crude oil in return for U.S.-made motor vehicles.

C483   "China Urges Japan to Double Two-Way Trading Target--Also Asked to Increase Crude Oil Imports," by A.E. Cullison. Journal of Commerce (Aug. 31, 1978):9, 31.
        A commentary on China's proposal to double the two-way trade target of the eight-year trade pact with Japan concluded in February 1978. The original pact envisaged a bilateral trade pact of $20 billion, with the export of oil by China gradually increased to 30 million tons by 1985. Now China has made a new proposal to double the total volume of trade to $40 billion and to increase the shipment of oil from the original target of 30 million tons to 50 million tons by 1985.

C484   "China's Oil Is Offered to America." <u>Journal of Commerce</u>
(Aug. 31, 1978):30.
    There are unconfirmed reports that Chinese crude oil has
been offered to the Gulf Oil Corp. and that some crude sold
to South Korea by Gulf actually comes from China.

C485   "Jenchin (Jenchiu)--A New Big Oilfield in China." <u>Economic</u>
<u>Reporter</u> (English supp.) 3 (July/Sept. 1978):38-39. illus.
    News report on the development of a high-yielding oil
field at Jench'iu on north China's central plain in Hopeh
province. The drilling commenced in 1975, and by 1976 many
wells that produce one thousand to several thousand tons a
day have been completed. The Jench'iu field's crude oil output was already 10% higher than its designed capacity in 1977,
and the first 8 months of this year saw its state plan over-
fulfilled.

C486   "Shantung--Nanking Oil Pipeline Laid." <u>Economic Reporter</u>
(English supp.) 3 (July/Sept. 1978):40-41. illus.
    A news report on the completion of a 1,000-kilometer trunk
pipeline from Linyi in Shantung province to Nanking harbor in
Kiangsu province. The building of this large-caliber pipeline
makes it possible for crude oil from Shengli in north China to
be shipped from the port of Nanking to the city of Shanghai,
Chekiang province, and Wuhan along the Yangtze River.

C487   "China Seeks Bids," by T.J. Stewart-Gordon. <u>World Oil</u> 187,
no. 4 (Sept. 1978):21. illus.
    Analyzes China's decision to seek bids from foreign firms
for offshore oil exploration. Although China is seeking help
from foreign concerns, the extent of the offer and the sort of
aid China wants are not clear. The form of payment to the
participating companies is also shrouded in mystery. Stewart-
Gordon notes that China seems to prefer paying for services in
cash rather than sharing oil, as desired by foreign oil
companies.

C488   "Closer Links with the Outside World." <u>Petroleum Economist</u>
45, no. 9 (Sept. 1978):393-94.
    Reports that China has indicated its readiness to cooper-
ate with Japan and other leading Western countries in oil
exploration.

C489   "China's Oil Workers Gird Up for New Task," by Po-hsi Fang.
<u>Peking Review</u> 21, no. 35 (Sept. 1, 1978):20-23. illus.
    An account of interviews with several model oil field
workers at the National Conference on Learning from Tach'ing
in Industry held at the Tach'ing complex. Among those highlighted are Tsai Kuo-chen, a political instructor for an oil
extracting team, and Lu Chu, a female worker at the Tach'ing
oil field.

C490   "China's Tough Oil Bargain," by James Strodes. <u>Far Eastern</u>
<u>Economic Review</u> 101 (Sept. 1, 1978):96-97.
Invitations which several American oil companies have re-
ceived to bid on China's offshore projects have raised much
speculation on the possible size of U.S. investment in these
ventures. The author warns that since the Chinese are very
tough bargainers, it is premature to predict the amount of
possible foreign investment in the project or even the likelihood of a joint venture. Strodes believes that the Chinese
will likely seek contracts calling for services and management
performed by foreign firms in return for fixed fees payable in
Chinese oil.

C491   "China Beckons the West--Watching the World," by Larry
Aulderidge. <u>Oil and Gas Journal</u> 76, no. 36 (Sept. 4, 1978):
56. illus.
China's pragmatic leaders want to modernize their country
by the turn of the century, and oil is regarded as a major
means for securing foreign currency to pay for imported technology and industrial equipment. Accordingly, China is en-
couraging Western prospectors to take part in developing her
promising offshore oil resources. Aulderidge notes that the
offshore area near Hainan Island shows good potential for
major U.S. oil concerns who wish to participate in joint ventures with the Chinese.

C492   "Isocyanates Unit Bought by China from Japan." <u>Chemical</u>
<u>Marketing Reporter</u> (Sept. 4, 1978):4.
An estimated $31 million contract has been awarded to a
Japanese consortium of three firms--Nippon Polyurethane
Industry Co., JGC Co., and C. Ito and Nishi Niho Trading Co.--
for the construction in Shantung of a 3-million-square-meter-
a-year synthetic leather plant. The plant will become operational in mid-1981.

C493   "New China Hand: Pennzoil for the Lamps of China?" <u>Forbes</u>
122, no. 5 (Sept. 4, 1978):90. illus.
A news report on Pennzoil president Huge Liedtke's con-
nections with China. Because of his friendship with George
Bush, former U.S. liaison chief to Peking, Liedtke seems to
have an edge over other oil majors in his access to Chinese
leadership, which might help him win the most coveted
contracts.

C494   "China Invites Mobil, Fifth U.S. Oil Firm, to Discuss Exploration." <u>Wall Street Journal</u> 45 (Sept. 5, 1978):17.
In addition to Pennzoil, Exxon, the Union Oil of California, and Phillips Petroleum, China is extending an invitation
to the Mobil Corp. to send a delegation to Peking for discus-
sions about the exploration of her offshore oil. The Mobil
delegation is expected to leave for Peking sometime this fall.

C495   "China Offers Oil for Technology," by Eric Bourne. <u>Christian Science Monitor</u> (Sept. 6, 1978):3.
       China is described as ready to trade its oil for the technology it needs to achieve economic modernization by the turn of this century. Rumania can supply expertise and industrial equipment for the Chinese petrochemical industry. Yugoslavia is also considered a possible partner for joint undertakings in the development of China's energy resources. These possibilities have been enhanced by the visit of China's party chairman Hua Kuo-feng to both countries in September 1978.

C496   "China Signs Pact with U.S. Firms for Oil Search," by A.E. Cullison. <u>Journal of Commerce</u> (Sept. 7, 1978):1, 17.
       This news report from Japan quotes China's Vice-Premier Teng Hsiao-p'ing as telling Japanese reporters that China has already signed agreements with U.S. oil corporations for joint exploration of China's offshore oil deposits. Neither the locations nor the names of corporations involved, however, were revealed. The news is regarded as highly upsetting to the Japanese, since they are the major buyers of Chinese crude oil and consider themselves as having better access to the Chinese on petroleum deals.

C497   "Reported Oil Accord of China, U.S. Firms Is Called Premature." <u>Wall Street Journal</u> (Sept. 7, 1978):14.
       Reports about an agreement between China and U.S. oil companies for the exploration of China's undersea oil deposits are called premature. Sources close to the oil companies consider the talks between China and the U.S. oil majors as only preliminary. The same sources suggest that the Japanese report on the U.S.-China accord on offshore exploration might be based on a misinterpretation of Vice-Premier Teng Hsiao-p'ing's remarks on such a possibility.

C498   "China Says It Prefers Japanese Partnership in Offshore Oil Projects." <u>Wall Street Journal</u> 49 (Sept. 11, 1978):20.
       Japan's Kyōdō News Service quotes Chinese Vice-Premier Kang Shi-en as telling Japanese reporters that China prefers Japan to others in the development of her offshore oil as long as the Japanese offer the same terms. The Japanese will presumably help China explore her offshore oil deposits in the Pohai Gulf, the Yellow Sea, and the South China Sea.

C499   "Another Super-Deep Well." <u>Peking Review</u> 21, no. 37 (Sept. 15, 1978):28.
       Peking reports on the drilling of an oil well in Szechuan province to a record depth of 7,175 meters. The previous record was set in 1977, when the same drilling team sunk a well to 7,058 meters in the same province. These successes represent the application of a newly developed drilling technique in China.

C500   "The Chinese Exchange," by W.A. Backman.   <u>Oil and Gas Journal</u>
        76, no. 38 (Sept. 18, 1978):85.
            There has been a heavy flow of commercial and technical
        traffic on petroleum business between China and the U.S. since
        early 1978. The visit of energy chief Schlesinger to China
        may well herald the start of Sino-American negotiations for
        the exchange of U.S. oil technology in return for Chinese
        crude oil. With the encouragement of the U.S. government,
        many U.S. oil companies are negotiating deals with China for
        the joint development of her offshore reserves.

C501   "ICI Gets Invitation to Participate in Taiwan in Plastic Ven-
        ture." <u>Chemical Marketing Reporter</u> (Sept. 18, 1978):5, 32.
            The government of Taiwan has invited the Imperial Chemical
        Industry of Great Britain to take part in a joint venture for
        the building of a 60,000-ton-a-year polypropylene plant at a
        cost of $100-120 million. A company decision is pending. The
        Imperial Chemical Industry already has invested $30 million in
        a methyl methacrylate plant in Taiwan that is slated to start
        production in 1978.

C502   "Polyester Facilities Mulled in China." <u>Chemical Marketing</u>
        <u>Reporter</u> (Sept. 18, 1978):7.
            China is planning the construction of two polyester plants
        capable of producing 530,000 tons and 180,000 tons a year.
        The total value of the project is estimated at $526 million.
        Kanebo, a textile manufacturer, and Hitachi, a large electri-
        cal firm, have been asked to submit bids for construction of
        the plants.

C503   "Xylene Rounds Out Taiwan Chemical Group." <u>Chemical Marketing</u>
        <u>Report</u> (Sept. 18, 1978):7.
            The Chinese Petroleum Corp. has completed Linyüan Petro-
        chemical complex, which will produce 60,000 tons of o-xylene
        and 200,000 tons of p-xylene to the downstream users in
        November 1978. The company is currently recruiting chemical
        groups in Taiwan for upstream construction at the complex.

C504   "Oil Federation Head Cautions Government Headlong Import of
        Chinese Oil." <u>Japan Economic Journal</u> (Sept. 19, 1978):7.
            Reports on warning issued by Masami Ishida, the president
        of the Petroleum Association of Japan, that Japanese oil re-
        fineries will be reluctant to accept Chinese crude if the
        price and quality of Chinese oil are not acceptable. The high
        cost of refining Chinese crude has made the Japanese wary of
        importing large quantities of waxy Chinese oil.

C505   "Japan-China Trade: Oil and Euphoria." <u>Economist</u> 268
        (Sept. 23, 1978):96-97.
            Japan counts on a flood of Chinese oil imports in the
        1980s in exchange for exports of technology and industrial
        equipment to China.

C506 "Hong Kong," by John Parke Wright. China Trade Report 14
(Oct. 1978):10-11. illus.
Wright discusses the key role Hong Kong is likely to play
as an outlet for the sale of China's petroleum products and as
a major export center for China's oil. He also speculates on
the possibility of Hong Kong's becoming a logistics base for
the developing onshore and offshore oil fields in the South
China region.

C507 "Methane from Manure." Petroleum Economist 45, no. 10 (Oct.
1978):437.
China is reported to produce bio-gas generated from
manure on a small scale, primarily for cooking at home.

C508 "China Seen Doubling Oil Production." Oil and Gas Journal 76,
no. 40 (Oct. 2, 1978):72-74.
Reports on a study of China's petroleum industry released
by the U.S. Bureau of Mines (entry A3, 1978 edition). Cites
the considerable achievements made by China's accelerated effort in oil exploration and drilling, with the addition of
many new producing oil fields such as Panshan and Jench'iu.
Significant increases have been achieved in producing natural
gas and in opening new gas fields. There also is a good
likelihood that crude oil output will double by late 1980.

C509 "Japan Peace Treaty's Effect on the Oil Industry," by George
Lauriat. Far Eastern Economic Review 101, no. 40 (Oct. 6,
1978):60-61.
The signing of peace treaty between China and Japan
creates both opportunities and problems for the two countries.
The positive factor is a steady and dependable supply of
Chinese crude to Japan, up to 40 million tons in 1985; the
negative factor is that treatment of waxy oil from Tach'ing
that can cost up to $10.5 billion in new cracking facilities
for Japan. Nevertheless, the treaty strengthens ties for
financing China's purchase of oil equipment from Japan and the
chances for joint development of China's offshore oil.

C510 "New Successes in Socialist Construction." Peking Review 21,
no. 40 (Oct. 6, 1978):18-21. illus.
Reports on key industrial achievements made in China so
far in 1978, including gains on petroleum fronts. Oil drilling is given a prominent place since its footage is up 77%
from the 1977 level. The production of crude oil and natural
gas has also made sizable gains. Among the most notable is
the completion of a new oil pipeline extending 1,000 kilometers from Hopeh province to the Naking harbor in Kiangsu
province. Construction of the pipeline started in October
1975 and was completed in July 1978.

C511   "Riding on China's Pendulum," by Susumu Awanohara. <u>Far</u> <u>Eastern Economic Review</u> 101, no. 40 (Oct. 6, 1978):85-86. illus.

        The importation of a urea plant for Tach'ing petrochemical complex illustrates a new swing of the political pendulum in China's economic decision-making. The xenophobic fever seems to be over and foreign expertise is respectable again. China has ordered a large ethylene plant from Japan to be built in the Tach'ing oil complex. But the author warns that unless China can make a rapid progress in its modernization, there is no assurance that the political pendulum might not swing back to the old rigidity.

C512   "Geosource Plans to Sell Oil Equipment to China." <u>Wall Street</u> <u>Journal</u> (Oct. 9, 1978):22.

        The Houston-based Geosources, Inc. has signed four contracts worth $15 million with the China National Oil and Gas Exploration and Development Corp. for geophysical equipment.

C513   "Soviet, Chinese Drilling Shows Different Emphasis." <u>Oil and</u> <u>Gas Journal</u> 76, no. 41 (Oct. 9, 1978):28.

        Reversing the earlier policy of independence and self-reliance on drilling, China is increasingly turning to the West for offshore technology and equipment. Recently, China has placed orders for several rigs built abroad, and is said to be considering some kind of joint venture with Japanese and U.S. oil companies for exploration of its vast offshore oil deposits in the Pohai Gulf and other areas.

C514   "Norway to Help China Oil Probe." <u>Japan Times</u> (International ed.) (Oct. 10, 1978):5.

        Norway's expertise may be used for China's development of offshore oil resources, with the possible supply of advanced oil equipment employed in Norway's offshore operations in the North Sea. Detailed discussion on this subject will be held when Norway's Energy Minister visits China in late 1978.

C515   "Sales of Chinese Oil to Power Firms." <u>Japan Economic Journal</u> (Oct. 10, 1978):7.

        A news report on the failure of Toshio Komoto, Japan's Minister of International Trade and Industry, to persuade Japan's power industry to boost imports of crude oil from China to reduce reliance on imports from other oil producing countries. It is expected that Komoto's shelving of his plan for Chinese oil will have some impact on the future of Sino-Japanese trade.

C516   "Jenchiu--A New High-Yielding Oilfield." Peking Review 21,
        no. 41 (Oct. 13, 1978):3-4. illus.
            A high-yielding oil field, with average oil wells produc-
        ing more than 1,000 tons a day, has been built at Jench'iu
        near Peking. It took only a year from the opening of the
        field to the production of oil. The high speed and the re-
        sults with which the oil field has been built is unprecedented
        in the history of China. After the oil find, tens of thou-
        sands of technicians and workers were mobilized to construct
        this field. A 200-kilometer pipeline, two big pumping sta-
        tions for crude oil, and other logistics projects have also
        been completed.

C517   "Science in China: A New Long March Begins," by Richard J.
        Seltzer. Chemical and Engineering News 42 (Oct. 16, 1978):
        34-68. color illus.
            The author's special reports on the current state of
        science and technology in China, with an emphasis on the
        chemical industry. A section of this essay deals with the
        development of petrochemical manufacturing, which is now said
        to be capable of adapting and modifying Western technology to
        China's domestic needs. The author particularly points to
        China's success in the operation of three oil shale plants,
        the only such plants in the world. Mention is also made of
        China's research in petroproteins from yeast based on oil.

C518   "Will There Be Plenty of Oil after All?" U.S. News and World
        Report 15 (Oct. 16, 1978):67-68. illus.
            The prospects of increased crude oil output in Canada,
        China, and Mexico are discussed. China's apparent decision
        to seek Western technology in developing the estimated 70
        billion barrels of oil locked in 46 potential oil basins can
        help ease world oil shortages in the 1980s.

C519   "Promise of Major Finds in China's Coastal Area Draws U.S.
        Oil Firms; Asian Bonanza?" by James Tanner. Wall Street
        Journal (Oct. 19, 1978):1, 34.
            A report on U.S. oil companies' interest in the develop-
        ment of China's offshore oil on the eve of the departure of
        Energy Secretary James Schlesinger to Peking. The article
        speculates on Schlesinger's mission, possible oil reserves in
        China's offshore areas, prospects for U.S. sales of oil equip-
        ment to China, the estimated costs of offshore explorations in
        China's continental shelf, and Chinese oil's prospective ex-
        port markets.

C520   "Chinese Drillers Strike Oil in Sinkiang Province," by Ian
        McKenzie. Japan Times (International ed.) (Oct. 23, 1978):5.
            News reports from Urumchi, the provincial capital of
        Sinkiang, indicate that a substantial oil find has been made

near the western edge of the Tarim basin between the towns of
Kashgar and Aksu. The oil is reported to be of high quality.

C521   "Chinese Petroleum Corp." Oil and Gas Journal 76, no. 43
        (Oct. 23, 1978):72.
            Taiwan's key oil refinery will undergo a major expansion
        at its plant site in Kaohsiung. The Chinese Petroleum Corp.'s
        project includes a fluid catalytic cracker, a computerized
        control Merox, sulphur recovery, and amine gas treating units.
        C-E Lummus won the contract with CPC worth $70 million. The
        project is due to be completed in 1980.

C522   "The Scramble to Exploit China's Oil Reserves." Business Week,
        no. 2658 (Oct. 30, 1978):155-56. map.
            Despite the absence of formal diplomatic ties with China,
        U.S. oil companies are among a dozen international oil con-
        cerns negotiating with China for the development of her vast
        offshore hydrocarbon deposits. There is a possibility that
        China might be ready for some formula of sharing oil with
        international oil majors. American oilmen are confident that
        in the deep water drilling technology, U.S. superiority will
        give them an edge over the others and believe that the day
        might not be too far off for the signing of the first Sino-
        American oil contract.

C523   "China Asks Petrocanada and Ranger Oil to Participate in Off-
        shore Exploration." Petroleum Economist 45, no. 11 (Nov.
        1978):489.
            Two major Canadian oil concerns, Petrocanada, Ltd. and
        Ranger Oil, Ltd., have been invited to send delegations to
        China to discuss possible Canadian assistance in the develop-
        ment of China's offshore oil.

C524   "China Completes Major Oilfield in Jenchiu." Petroleum
        Economist 45, no. 11 (Nov. 1978):487.
            Quotes Chinese sources on the development of a high-
        yielding oil field at Jench'iu in Hopeh province. The field
        is said to have many wells producing crude oil at the rate of
        several thousand tons a day.

C525   "China Estimates Offshore Potential at 10-Billion Tons of
        Crude." Petroleum Economist 45, no. 11 (Nov. 1978):487.
            Sources in China have disclosed that China's probable
        offshore hydrocarbon deposits are at least in the range of
        10 billion tons.

C526   "Oil and Gas in China (III)," by R.W. Scott. World Oil 187,
        no. 6 (Nov. 1978):69-72. color illus.
            Reports on a second visit to the Tach'ing oil field and
        offshore operations in the Pohai Gulf by the editor of World
        Oil, with more than a dozen colorful illustrations.

C527   "Pipeline for Shipping Crude from Shengli Oil Fields to
       Nanking Harbor Completed." Petroleum Economist 45, no. 11
       (Nov. 1978):478.
           A major trunkline connecting Shengli and the oil terminal
       at Nanking Harbor has gone on line. The pipeline starts at
       Linyi in Shantung province and ends at the Yangtze River port,
       running over 1,000 kilometers.

C528   "Southern Asia: China Prepares for Massive Oil and Gas Devel-
       opment." Pipeline Industry 49 (Nov. 1978):73-74. illus.,
       table.
           China is engaged in the extensive construction of a crude
       oil trunkline and gas transmission line totaling more than
       3,300 miles and scheduled for completion in 1980.

C529   "Taiwan Energy Plans Emphasis Electricity," by Vaclav Smil.
       Energy International 15, no. 11 (Nov. 1978):49-51. illus.,
       map.
           The rapid development of Taiwan's economy has been accom-
       panied by a huge increase in the consumption of energy. So
       far, crude oil from the Middle East has been the major source
       of energy in this island country. With the domestic output of
       oil accounting for a fraction of her needs and the cost of
       imported oil rising rapidly, Taiwan's government is shifting
       from oil-fired power generation to nuclear power generation,
       as attested by the construction of a group of large nuclear
       power plants. The first and second went on line in 1977 and
       1978.

C530   "Japanese Firms Face Tough Problems in Using Chinese Oil,
       Coal." Business Asia 44 (Nov. 3, 1978):345-46.
           The Japanese petroleum refining industry is lukewarm
       toward increased Chinese oil imports. The power industry in
       particular has been very reluctant to use Chinese crude be-
       cause of its wax content, which makes refining expensive.
       Thus the only chance for increased imports hinges upon the
       Japanese government's becoming involved in refining Chinese
       oil.

C531   "Sobering Thought in the Oil Rush," by Melinda Liu. Far
       Eastern Economic Review 101, no. 44 (Nov. 3, 1978):45. map.
           A sobering note to many international oil executives who
       have been euphoric over the prospects of getting Chinese oil
       contracts. Despite the scramble for Chinese oil, there has
       been no concrete information on China's oil reserves and pro-
       duction. Liu's own estimate is between 116 and 127 million
       tons by 1980, of which only 20 million tons will be available
       for export.

C532   "China Boasts Record Drilling, Production, New Oil Line." Oil
and Gas Journal 76, no. 45 (Nov. 6, 1978):24-25. illus.
        Reports from China indicate that oil and gas production
was boosted by 11% and 14%, respectively, during the first
8 months of 1978. Drilling footage is reportedly up 77% in
the same period, with 88% of the 1978 state plan completed by
August 1978. China also has completed a north-south oil pipeline with a 621-mile trunkline from the Shengli oil field in
Shantung province to the oil terminal at Nanking in Kiangsu.
China is also said to be capable of building a 2.5-million-ton
refining facility by herself.

C533   "China's Crash Jenchiu Field Development Nets High Flows."
Oil and Gas Journal 76, no. 45 (Nov. 6, 1978):24-25.
        Liberally quoting an article from Peking Review (entry
C516) and a study by the U.S. Bureau of Mines (entry A3,
1978 ed.), the Journal describes the crash development of
China's new oil field at Jench'iu. Tens of thousands of
workers have been mobilized to build the oil field and put it
into production within a year. An average well is said to
produce more than 7,300 b/d of crude oil, and the output from
only a dozen wells at Jench'iu is sufficient to satisfy the
Peking refinery, one of the largest in China.

C534   "Russian Press Attacks U.S. Role in China's Oil Buildup." Oil
and Gas Journal 76, no. 45 (Nov. 6, 1978):111.
        The Soviet Union is accusing the U.S. of trying to help
build up the petroleum industry of China while attempting to
slow down the development of oil production in the USSR by
refusing to sell the latter the same oil equipment that it
sells China. The party organ Pravda, the government paper
Izvestia, and the economic weekly Ekonomicheskaya Gazeta all
parrot the allegation of Peking-Washington connection on oil,
expressing disapproval of U.S. help to China's oil industry
and China's use of oil for diplomatic objectives.

C535   "Who Is Making the Big Business Deals with China?" Business
Week, no. 2559 (Nov. 6, 1978):76-77.
        Reports that foreign firms are lining up to sell capital
goods and technology to China, and that oil is considered the
biggest prize. Currently, in addition to four American international oil companies, there are British, Dutch, Italian, and
Japanese oil firms in various stages of negotiations with the
Chinese for the development of her offshore oil deposits.
China has ordered $500 million worth of oil equipment, but she
is expected to purchase more oil technology and equipment from
the U.S. and other Western countries.

C536   "Kanematsu-Gosho Purchases Light Oil from China." <u>Japan Economic Journal</u> (Nov. 7, 1978):6.
     Kanematsu-Gosho, Ltd., a major Japanese trading company, has imported 16,000 tons of Chinese light crude at a price of $17.40 per barrel. The company plans to sell half of the crude in December and the balance in early 1979. The company is also contemplating the importation of light oil from China on a long-term basis.

C537   "China Agrees to Sell 634,900 Barrels of Oil to Refinery in Italy." <u>Wall Street Journal</u> (Nov. 8, 1978):38.
     A report from Rome indicates that China has made its first sale of crude oil to a West European country. The Garrone Refinery in Italy will receive its first shipment of 317,450 barrels of Chinese crude in December. A second delivery will be due in January 1979. The Chinese oil will be tested to determine its commercial use. The price of this sale has not been disclosed.

C538   "Sino-Japanese Oil Drilling Will Be Tried in Pohai Gulf." <u>Japan Economic Journal</u> (Nov. 14, 1978):8.
     According to a report released by the Japan National Oil Corp., an agreement has been reached between China and Japan to jointly explore a 20,000-square-kilometer concession in the southern part of the Pohai Gulf. Japan will provide drilling rigs, development planning, and production supervision, while the Chinese will take charge of training personnel and procuring materials and offshore drilling equipment. The Japanese also expect China to agree to a joint venture in developing undersea oil deposits off the mouth of Pearl River in the South China Sea.

C539   "Last Untapped Pools: Late-Comer China Emerging as Giant Petroleum Power." <u>Journal of Commerce</u>, no. 24337 (Nov. 16, 1978):7.
     A commentary on China's oil in the wake of Energy Secretary Schlesinger's three-week visit to China. According to Schelsinger, China's oil depsoits amount to at least 100 barrels--3 times the proven reserves of the U.S. China has also set a target of crude oil production at 4-6 million b/d by 1985. This is an ambitious goal in view of China's current crude oil output, estimated at 1.8 million b/d in 1977.

C540   "Coastal State Gas Signs an Agreement to Buy Chinese Oil." <u>Wall Street Journal</u> (Nov. 22, 1978):20.
     Coastal State Gas Corp., a major independent oil and gas company, will become the first U.S. firm to import crude oil from China. Coastal State Gas has signed an accord in Peking with China's state agency, the China National Chemical Import and Export Corp., for the purchase of 3.6 million barrels of crude. These are to be shipped to the company's refinery at

Hercules, California. The first shipment will take place early in 1979, with the last shipment being made by the middle of the year. The crude oil is believed to be produced at the Tach'ing oil field and has a high paraffin content, but is low in sulphur.

C541   "Peking to Sell Oil Directly to U.S. Company," by J.P. Smith. Washington Post (Nov. 22, 1978):A1, A12.
        A contract has been signed for the sale of China's Tach'ing crude oil to Coastal State Gas Corp.

C542   "U.S. to Get Low-Sulfur Chinese Oil--Coastal States Signs Accord for Imports," by William K. Stevens. New York Times (Nov. 22, 1978):D1, 3.
        The Coastal States Gas Corp. has become the first U.S. oil company to import Chinese crude oil for domestic use. Under the accord with the Chinese, 3.6 million barrels of low-sulphur crude oil valued at nearly $50 million will be unloaded at the company's refinery at Hercules, California, where Chinese crude will be mixed with other petroleum and refined as fuel oil. Shipments will begin early next year and last for about six months. The deal with China is considered by the Departments of State and Energy to signal closer energy ties between the two countries.

C543   "New Oil Refineries Completed in China." Economic Reporter (English supp.) 4 (Oct./Dec. 1978):30.
        Mention is made of three oil refineries going on line during 1978: the Chekiang Oil Refinery in Chekiang province, the First Petrochemical Plant in Sinkiang province, and the Wuhan Petrochemical Plant in Hupeh province. Chekiang has an annual processing capacity of 2.5 million tons of crude oil, turning out gasoline, kerosene, diesel oil, heavy oil, and liquefied petroleum gas.

C544   "Taching/Pohai Journal," by Stephanie R. Green. China Business Review 5, no. 6 (Nov./Dec. 1978):10-20. illus., tables.
        Eyewitness reports on the Tach'ing oil base and the Pohai Gulf offshore oil field during a three-week tour of China in September 1978 by an American mission. Describes managerial organization, exploration, drilling, production, and refining at the Tach'ing oil complex; also gives an account of China's offshore operations in Pohai Gulf, including a visit to a Chinese-made offshore drilling rig and an inspection tour of China's oil port facilities and the Talien Shipyard, where China's own jackup rigs were under construction.

C545 "China's Energy Plans Call for Foreign Assistance," by Vaclav
      Smil. Energy International 15, no. 12 (Dec. 1978):27-29.
      illus.
            Smil critically analyzes China's energy plans for the
      years 1978-1985, disclosed at China's National People's Con-
      gress in February 1978. He contends that China's petroleum
      technology is 10-20 years behind the West in its sophistica-
      tion and quality. Her offshore technology, oil processing,
      oil pipe manufacturing, and shipbuilding for oil tankers need
      upgrading with advanced technology from the West. Smil con-
      cludes that China needs huge capital investments and substan-
      tial technology transfers from the advanced countries in order
      to modernize her energy industry.

C546 "Geophysics in China," by T.J. Stewart-Gordon. World Oil 187,
      no. 7 (Dec. 1978):21.
            Deals with China's geological characteristics and pros-
      pects for oil exploration.

C547 "Oil and Gas in China (IV)," by Robert W. Scott. World Oil
      187, no. 7 (Dec. 1978):41-52. color illus., maps.
            A colorful description of the World Oil editor's 1978
      visit to China's major oil terminal, offshore drilling
      operations, oil equipment manufacturing facilities, and the
      Tach'ing oil base, including discussion on China's offshore
      technology and oil producing industry.

C548 "Toward Worldwide Oil Sales." Petroleum Economist 45, no. 12
      (Dec. 1978):531.
            Recent oil sales contracts with Coastal State Gas Corp.
      of the U.S., Tecnotrade of Italy, and the barter deal with
      Brazil seem to augur well for China to become a worldwide
      crude oil exporter, even though the quantities are still very
      modest.

C549 "Ore for Oil--Brazil-China Trade." Economist, no. 7057
      (Dec. 2, 1978):86, 89.
            Reports on the accord reached in Peking between China and
      Brazil on trading Chinese crude oil for Brazilian iron ore and
      pig iron. The barter deals call for the shipment of 20,000
      b/d of Chinese crude to Brazil in 1979 in exchange for 2.5
      million tons of Brazilian iron ore and 200,000 tons of pig
      iron. The export of Chinese crude to Brazil is to rise to
      30,000 b/d by 1980. In the years to come, Brazil is expected
      to export increasingly large quantities of iron ore to China
      and in return she expects to buy from China all the oil she
      wants to sell.

C550   "Chinese Exports to Hong Kong Show Big Gain, Centering on Oil
Products." <u>Japan Economic Journal</u> 830 (Dec. 3, 1978):4.
      Reports on China's expanding share of Hong Kong's petro-
leum products market. Statistics show that during the first
9 months of 1978, these exports registered a 67% gain over
their 1977 level. China has thereby increased her share to
25%, accounting for $100 million in sales in the first 3 quar-
ters of 1978.

C551   "Now, It's Chinese Oil for Lamps of the U.S." <u>Chemical Week</u>
123, no. 23 (Dec. 6, 1978):22.
      Reports on China's sale of her crude oil to the Coastal
State Gas Corp. through the good offices of Julian Sobin, the
chairman of Sobin Chemicals in Boston. The oil deal, for
about 3.6 million barrels to be shipped within 6 months, is
valued at about $50 million. The independent gas company is
going to process the Chinese crude for low sulphur fuel oil at
the company's refinery at Hercules, California, which is now
using Indonesian crude almost exclusively.

C552   "Highway-Pipeline Double-Utility Bridge." <u>Peking Review</u> 21,
no. 49 (Dec. 8, 1978):31.
      Announces the completion of a 1,922-meter highway bridge
incorporating an oil pipeline across the Huai River in
Kiangsu province. The bridge completes a segment of the
Shantung-Nanking pipeline carrying crude oil from the north to
the port of Nanking for shipment to southern China. The pipeline is 720 millimeters in diameter and is supported by light
steel props on one side of the bridge.

C553   "The Shanghai Petrochemical Complex," by Mao Lin. <u>Peking
Review</u> 21, no. 49 (Dec. 8, 1978):10-12. illus.
      An up-to-date sketch of the vast Shanghai General Petro-
chemical Works in Chinshan county, 70 kilometers southwest of
Shanghai. The first phase of construction, with eighteen
processing units (half of them imported) and ancillary facilities, has been completed. The site of the complex was re-
claimed from the sea in 1972, and construction commenced in
1974. Trial operations began in 1976 and the whole complex
went on line in 1977. In addition to refining crude oil, the
petrochemical plants are turning out 100,000 tons of vinylon,
acrylics, and terlene annually. About 50,000 professional
staff, workers, and their family members live in the compound
of this petrochemical complex.

C554   "China Official Spells Out Trade Policies, Says U.S. Gets
Equal Footing with Others." <u>Wall Street Journal</u> (Dec. 19,
1978):21.
      Reports on China's new trade policy toward the U.S. as
enunciated by Li Chiang, China's Foreign Trade Minister. Li
indicated that China is ready to pay in oil for the services

rendered by foreign companies who help develop her offshore
oil. He told reporters that there would be opportunities for
many foreign oil firms to take part in the exploitation of
China's vast offshore oil.

C555   "China Looks Outward for CIP Buildup." Chemical Week 123,
        no. 25 (Dec. 20, 1978):11-13. illus., table.
            The expansion of China's chemical processing industries as
        part of her economic modernization plan provides an enormous
        future market for Japanese and Western chemical and petro-
        chemical plants and facilities. This article provides infor-
        mation about deals China has made with Western countries and
        Japan for the purchase and construction of chemical and petro-
        chemical plants before and after 1978. The Chinese market is
        regarded as substantial at present and as offering great
        promise in the future.

C556   "Oriental Express: Small Firms to Benefit from New China
        Trade along with Big Ones," by Charles W. Stevens. Wall
        Street Journal (Dec. 20, 1978):1, 28.
            Greerco, Inc., a small oil processing machinery maker with
        an annual sale of $6 million, has succeeded in obtaining an
        order for 10 wax molders, valued at $6.8 million, over the
        next two years. It has been an arduous process of negotia-
        tions which required the utmost patience on the part of the
        seller. The machines will be used by oil refineries in China
        to process paraffin refined from crude oil and mold it into
        11-pound blocks for industrial use.

C557   "China Agrees to Buy Seven Drilling Rigs from Unit of LTV."
        Wall Street Journal (Dec. 21, 1978):10.
            Reports on an accord between China and Continental Emsco,
        a subsidiary of LTV Corp., for the purchase of 7 onshore and
        offshore drilling rigs valued at $40 million. This is the
        first deal ever between China and U.S. companies for the sale
        of land rigs. Of the seven rigs, two are 1,500-horsepower
        diesel-powered land rigs. The other five are 2,000-horsepower
        electric rigs for use on offshore drilling platforms.

C558   "China Buys 7 Oil Rigs." New York Times (Dec. 21, 1978):D4.
            Reports on China's purchase of seven oil drilling rigs for
        use in onshore and offshore exploration. The builder of the
        rigs is Continental-Emsco Co., unit of the LTV Corp. of
        Dallas, Texas. This marks China's first acquisition of land
        rigs from the U.S. Valued at $40 million, they will be
        delivered in 1979 and 1980.

C559   "Meet the Mrs." <u>Forbes</u> 122, no. 13 (Dec. 25, 1978):61-62.
       illus.
            A brief inside story of the first sale of China's crude
       oil to the U.S. The deal for 500,000 tons of Chinese crude
       with Coastal State Gas Corp. was arranged by Julian Sobin and
       his wife Lee, owners of the medium-sized Sobin Chemicals,
       based in Boston. It is through their trust and friendship
       with the Chinese officials, that the accord was reached.

C560   "China Awards Pullman Kellogg Pact for Petrochemical Facility."
       <u>Journal of Commerce</u> (Dec. 27, 1978):5.
            Pullman Kellogg, a division of Pullman, Inc., has been
       awarded a contract worth millions of dollars by the Chinese
       government to provide advisory services in engineering, pro-
       curement, and construction for large petrochemical facilities
       using technology developed by Hercules and UOP of the U.S.
       Early in 1973, Kellogg was also awarded contracts to build
       eight fertilizer plants in China. Six of these are already on
       line and the remainder are scheduled for completion in 1979.

C561   "Pullman Division Gets Chinese Job for Work on Petrochemical
       Unit." <u>Wall Street Journal</u> (Dec. 27, 1978):5.
            Pullman Kellogg, a subsidiary of Pullman, Inc., announced
       that it has been awarded a multi-million dollar contract by
       the Chinese government to provide advisory services for the
       engineering, procurement, and construction of a large petro-
       chemical complex. The Chinese facility will employ technology
       developed by Hercules and UOP, both U.S.-based companies.

C562   "Pullman Unit Set to Aid Major Chinese Oil Project." <u>New York
       Times</u> (Dec. 27, 1978):D2.
            Reports on a petrochemical contract awarded by the Chinese
       government to Pullman Kellogg, a division of Pullman, Inc.,
       for help in building a large petrochemical facility in China.
       The facility consists of processing plants to produce meta-
       cresol, butyl hydroxytoluene, and acetone. The contract,
       which is worth millions of dollars, will use technology devel-
       oped by Hercules and UOP. Pullman, Inc. is to furnish advis-
       ory services on the engineering, procurement, and construction
       of these petrochemical plants.

C563   "Group of Japan Firms Will Export to China 4 Petrochemical
       Plants." <u>Wall Street Journal</u> 126 (Dec. 29, 1978):15.
            The China National Technical Import Corp. has awarded a
       group of four Japanese companies contracts to build 4 petro-
       chemical plants valued at $205 million. The Chinese orders
       include 2 low-density polyethylene plants, each with an annual
       capacity of 140,000 tons, a high-purity terephthalic acid
       plant with an annual capacity of 225,000 tons, and a phenol
       and acetone plant capable of producing 50,000 tons a year.
       The Japanese concerns comprise 2 trading firms, C. Ito and

the Kosho Corp., and 2 industrial firms, Mitsui Petrochemical Industries and the Mitsui Engineering and Shipbuilding Co.

C564 "The Output of Crude Oil in the First Half of This Year Reaches the Highest Level in History." Chinese Economic Studies 12, nos. 1-2 (Fall-Winter 1978-79):139-41.
Translation of a report by New China News Agency on a 10.6% gain in crude oil production in the first half of 1977 over that of 1976. The story was carried in Jen-min jih-pao (People's Daily) on July 10, 1977.

C565 "China's Potential Still a Guessing Game," by A.A. Meyerhoff. Offshore 39, no. 1 (Jan. 1979):48-56. illus., maps (some color).
A realistic assessment of China's offshore oil reserves through information provided by an authority on China's petroleum industry.

C566 "Chinese Officials Ponder Next Move," by L. Leblanc. Offshore 39 (Jan. 1979):48-53. illus., map.
Analyzes China's official thinking on the development of her offshore oil reserves and the chances for foreign participation.

C567 "China to Buy U.S. Rigs, Chem Services." Oil and Gas Journal 77, no. 1 (Jan. 1, 1979):38.
Two U.S. firms, Continental-Emsco and Pullman Kellogg, were awarded contracts by China to build land rigs and render services on the construction of petrochemical facilities in China.

C568 "Japan Refinery Seeks Big Funds--Has Eye on Chinese Oil," by A.E. Cullison. Journal of Commerce (Jan. 3, 1979):1, 30.
A major Japanese oil company, Idemitsu Kosan, is planning to build a large refinery to process crude oil imported from China.

C569 "Chinese Offshore Oil Eyed--U.S., China, Japan May Cooperate," by A.E. Cullison. Journal of Commerce (Jan. 8, 1979):1, 32.
Reports on negotiations under way in Japan on the exploration of Phohai (Bohai) Gulf by China, Japan, and the U.S.

C570 "Three Japanese Groups Get Jobs from China Totaling $700 Million." Wall Street Journal (Jan. 9, 1979):17.
Contracts worth more than $700 million have been awarded to 3 groups of Japanese contractors for 4 ethylene plants, 2 of which will have a capacity of 300,000 tons a year.

C571   "China Plant Contract to German Concern." <u>New York Times</u>
        (Jan. 11, 1979):D9.
        A West German company, Lurgi Gesellschaften, received a
        contract valued at $500 million from China to build 3 petro-
        chemical plants. In addition, the company is said to have
        received contracts from China in December 1978 for the build-
        ing of chemical plants worth $750 million.

C572   "Lurgi Gets Contract Valued at $540 million for 5 Plants in
        in China." <u>Wall Street Journal</u> (Jan. 11, 1979):37.
        Lurgi Kohle & Mineraloeltechnik G.m.b.H. of West Germany
        received a contract for 5 petrochemical plants worth $540 mil-
        lion from China's National Technical Import Corp.

C573   "Chinese Moving to Expand Oil Industry Trade." <u>Oil and Gas
        Journal</u> 77, no. 4 (Jan. 22, 1979):24-25.
        To expand her oil industry, China seeks help from Japan in
        developing offshore oil from Pohai (Bohai) Gulf, also the U.S.,
        Canadian, and German aid in petrochemical technology.

C574   "The U.S. Discovers China," by W.A. Backman. <u>Oil and Gas
        Journal</u> 77, no. 4 (Jan. 22, 1979):39.
        Comments on business benefits the U.S. will gain from
        diplomatic recognition of China. The petroleum-related indus-
        tries in the U.S. are likely to be great beneficiaries of this
        move.

C575   "U.S. Oil Concerns Step Up Drive to Prod China into Reaching
        Accord on Drilling," by Barry Kramer. <u>Wall Street Journal</u>
        (Jan. 25, 1979):7.
        Reports on American oil companies' efforts to convince the
        Chinese to let them take part in the exploration of China's
        offshore oil deposits.

C576   "China Lets More Services, Equipment Contracts." <u>Oil and Gas
        Journal</u> 77, no. 5 (Jan. 29, 1979):96.
        A number of firms representing U.S., Japan, and West
        Germany have signed contracts with China to provide China oil
        equipment and petrochemical facilities.

C577   "Lamps of China to Get More Oil, but Not U.S." <u>National
        Petroleum News</u> 70 (Feb. 1979):36.
        Discusses the impact of the U.S. diplomatic recognition of
        Peking on the prospects for crude oil imports from China and
        the repercussions on oil markets in the U.S.

C578   "China Expanding Purchases of Oil Equipment, Technology," by
        Bob Tippee. <u>Oil and Gas Journal</u> 77, no. 7 (Feb. 12, 1979):
        36-37. illus.
        China has greatly increased its acquisition of oil equip-
        ment and technology from the Western countries. The shopping

list includes drilling rigs, offshore platforms, and seismic equipment.

C579   "China's Big Shopping List to Test U.S. Trade Policy," by
        W.A. Backman. Oil and Gas Journal 77, no. 7 (Feb. 12, 1979):
        35, 38-39. illus.
            Vice Premier Teng Hsiao-p'ing's visit to the U.S. opens up
        broad vistas for the expansion of Sino-U.S. trade, with U.S.
        oil technology to be exchanged for China's oil.

C580   "Soviets Hit China Oil Potential Reports." Oil and Gas
        Journal 77, no. 7 (Feb. 12, 1979):40-41.
            The Soviet Union questions the validity of oil data
        released by China and regards the West's assessment of China's
        oil potential as too optimistic.

C581   "China Oil Business Awaited by Houston," by William K.
        Stevens. New York Times (Feb. 14, 1979):D1, D8. illus.
            Vice-Premier Teng Hsiao-p'ing's visit to Houston has
        stirred great hopes for oil equipment manufacturers for a
        promising Chinese market in oil equipment.

C582   "British Petroleum to Conduct a Survey in China's Yellow Sea."
        Wall Street Journal (Feb. 16, 1979):14.
            Reports on the signing of letter of intent between China
        and British Petroleum for the firm to conduct a seismic survey
        in a designated area in the Chinese waters of the southern
        part of the Yellow Sea.

C583   "China Oil Deal with Japanese." New York Times (Feb. 16,
        1979):D3.
            A basic accord has been reached on Sino-Japanese explora-
        tion of undersea oil in the shallow waters of the Pohai
        (Bohai) Gulf at an estimated cost of 2 billion dollars.

C584   "China Learns from Norway," by Harold Munthe-Kaas. Far
        Eastern Economic Review 106, no. 8 (Feb. 23, 1979):96.
            Discusses the prospect of Sino-Norwegian cooperation on
        offshore technology and offshore leases.

C585   "Adventures of the 'Gang of Seventeen' in China," by George
        McCourt. Fortune 99 (Feb. 26, 1979):34-35.
            Details the negotiations conducted between Phillips
        Petroleum and its Chinese counterpart on the firm's partici-
        pation in the development of China's offshore oil.

C586   "Doing Business with China," by Roger Vielvoye. Oil and Gas
        Journal 77, no. 9 (Feb. 26, 1979):61.
            Comments on the success of British Petroleum's contract
        negotiations with Chinese authorities on the firm's explora-
        tion of the southern part of the Yellow Sea.

C587   "Financial Snags Are Seen Allowing China in Buying Japan's
        Steel, Oil Technology," by Eduardo Lachia. Wall Street
        Journal (Feb. 28, 1979):18.
            China's financial problems have caused Japanese trading
        company executives to worry about delays in China's purchase
        of Japanese oil technology.

C588   "China's Oilfields Train Technical Personnel," by Shu Yao.
        Economic Reporter (English supp.) 1 (Jan./Mar. 1979):39-40.
        illus.
            Describes spare-time vocational and refresher training
        classes conducted for workers and technicians at the Tach'ing
        (Daqing) oil field.

C589   "Geology of Gu-dao Oil Field and Adjacent Areas," by Z. Chen
        Si and Wang Ping. In Annual American Association of Petroleum
        Geologists SEPM Meeting (Houston, Apr. 1-4, 1979) Papers.
        Abstracted in Bulletin of the American Association of Petro-
        leum Geologists 63, no. 3 (Mar. 1979):524.
            The first revelation of an existing oil field, Kutao
        (Gudao), and its adjacent geological structure in the Pohai
        (Bohai) Gulf by Chinese geologists attending AAPG's annual
        meeting in Houston. (Abstract only available.)

C590   "Oil and Gas in the People's Republic of China," by R.W.
        Scott. Annual American Association of Petroleum Geologists
        SEPM Meeting (Houston, Apr. 1-4, 1979) Papers. Abstracted
        in Bulletin of the American Association of Petroleum Geolo-
        gists 63, no. 3 (Mar. 1979):524.
            An account on the state of the art of China's petroleum
        industry based upon the author's visits to the Tach'ing
        (Daqing) and Shengli oil fields, as well as his extensive
        discussions with specialists on China's oil industry.
        (Abstract only available.)

C591   "The Geologic of China's Oil." Economist 270 (Mar. 3, 1979):
        100-2. illus., maps.
            An extensive, up-to-date analysis and survey of China's
        demand and supply of petroleum resources, with detailed dis-
        section of her onshore and offshore prospects and export
        potential.

C592   "Chinese Sell Fuel to U.S. Company," by Anthony J. Parisi.
        New York Times (Mar. 15, 1979):D15.
            Reports on the sale of Chinese petroleum products to a
        small trading company, Chem-Oil Industries, in New York.

C593   "Key Projects under Construction." Beijing Review 22, no. 12
        (Mar. 23, 1979):13-14. illus.
            Describes the prospecting and development of oil bases in
        north China, such as the Takang (Dagang) and Jench'iu (Renqiu)
        oil fields.

C594   "Energy and Power." China Trade Report 17 (Apr. 1979):4-5.
        illus.
              Discusses the possibility of Sino-American cooperation in
        the exploration and development of China's petroleum resources
        in the aftermath of visits to China by the Energy Secretary
        Schlesinger.

C595   "International Report: People's Republic of China." Ocean
        Industry 14, no. 4 (Apr. 1979):411-13. illus. (some color),
        map.
              A report on the development of oil and gas in China in
        1979, consisting of four sections: (1) offshore development
        and exploration; (2) domestic construction of jackup rigs in
        Chinese shipyards; (3) offshore loading facilities; and
        (4) negotiations with Western oil firms for the development of
        China's petroleum resources.

C596   "China Pushing Expansion of Oil and Gas." Oil and Gas Journal
        77, no. 17 (Apr. 13, 1979):26-28. illus., map.
              Reports that China gives priority to increased petroleum
        productive capacity over the long term rather than current
        gains in the output of oil.

C597   "Offshore Claims Complicate Chinese, Vietnamese Talks." Oil
        and Gas Journal 77, no. 17 (Apr. 13, 1979):28.
              Analyzes Sino-Vietnamese dispute on offshore rights in the
        Gulf of Tonkin.

C598   "Oil Exploration and Equipment Purchases Remain Chinese
        Emphasis." Business China (Apr. 25, 1979):59-60.
              Reports that sales of oil equipment and technology to
        China by foreign firms in early 1979 remained very brisk
        despite signs of a slowdown in China's import of foreign
        goods.

C599   "Energy in China: Interview with Industry Officials."
        Petroleum News--Southeast Asia (Hong Kong) 10, no. 2 (May
        1979):10-11.
              An overall evaluation of China's current energy situation
        on the basis of an interview with Chinese energy officials.
        Stresses the steady increase of crude oil production since
        1949 through extensive exploration.

C600   "Russia Steps Up Anti-China Campaign." Offshore 39 (May
        1979):330-31, 333.
              The mass media in the U.S.S.R. have mounted an anti-China
        offensive, trying to discredit China's offshore oil potential
        and undercut her relationship with neighboring countries.

C601   "Coastal to Hike Chinese Oil Purchase." <u>Oil and Gas Journal</u>
        77, no. 21 (May 21, 1979):42.
            Coastal State Gas Co. of the U.S. has doubled its purchase
        of crude oil, diesel fuel, and gasoline from China.

C602   "China's Perplexing Energy Triangle," by Vaclav Smil.   <u>Energy</u>
        <u>International</u> 16, no. 6 (June 1979):25–27.   map.
            Evaluates China's energy supply picture, with emphasis on
        her potential for the export of crude oil.

C603   "Yumen Oil Field Plays Its Part," by Jian Qiu.   <u>China Recon-</u>
        <u>structs</u> 28, no. 6 (June 1979):42–43.   illus.
            A profile of the development of this 40–year–old oil field
        in China from its primitive state in 1938 to a relatively
        advanced stage in 1979.

C604   "China Expects Oil-Exploration Pacts," by Clyde H. Farnsworth.
        <u>New York Times</u> (June 4, 1979):D1+.
            China will soon permit American oil companies to conduct
        geophysical prospecting in China's offshore areas.

C605   "Exxon Unit Contracts with China to Conduct Offshore Examina-
        tion." <u>Wall Street Journal</u> (June 6, 1979):47.
            Esso Exploration, an Exxon subsidiary, has been granted a
        permit by China to conduct seismic surveys in a designated
        area in the South China Sea.

C606   "China Opens Offshore Oil Bidding––Within Limits," by Ron
        Scherer. <u>Christian Science Monitor</u> (June 7, 1979):11.   illus.
            An account of an accord signed by Exxon and Mobil with
        China on conducting seismic surveys of offshore oil in the
        South China Sea.

C607   "Mobil Corp. Unit Gets a Chinese Contract for Seismic Surveys."
        <u>Wall Street Journal</u> (June 7, 1979):5.
            Mobil Oil has signed a seismic survey contract with the
        visiting Chinese oil delegation to make geophysical examina-
        tions in a designated area in the South China Sea.

C608   "Bethlehem Delivers Jack-Up to China." <u>Oil and Gas Journal</u>
        77, no. 24 (June 11, 1979):29.   illus.
            Reports that Bethlehem Singapore, Ltd. delivered an off-
        shore oil rig including a platform to China. It will be used
        for oil exploration in the Pohai (Bohai) Gulf area.

C609   "BP Search in First Phase to Yellow Sea Oil Auction," by
        A. Chuter. <u>Engineer</u> 248 (June 14, 1979):14.
            Reports on British Petroleum's participation in the seis-
        mic survey for oil and gas deposits in China's Yellow Sea.

C610   "Taiwan Industries Feeling Oil Crunch," by A. Koffmann
O'Reilly. <u>Journal of Commerce</u> (June 19, 1979):10.
    The industrial community in Taiwan has been notified by
the Chinese Petroleum Corp., the state-owned oil agency, that
their demand for increases in imported foreign oil could not
be met.

C611   "Worldwide Drilling and Production: China Abounds in Contracts, Projects." <u>Offshore</u> 39, no. 7 (June 20, 1979):194.
table.
    China plans a major expansion into the development of her
offshore oil, with more contracts, international accords, and
joint ventures with Western oil companies in the offing.

C612   "Firms Poised for China Seismic Surveys." <u>Oil and Gas Journal</u>
77, no. 26 (June 25, 1979):26. map.
    Eight major international oil companies have been given
rights to conduct seismic surveys of China's continental
shelf, which has a total area of 250,000 square kilometers.

C613   "U.S., Chinese Meet to Swap Oil, Gas Reserves Ideas," by Lynn
Brenner. <u>Journal of Commerce</u> (July 5, 1979):1, 30.
    A brief description of the state of China's natural gas
industry according to John Kean, Chairman of the American Gas
Association, after his recent visit to China.

C614   "China Reveals First Detailed Oil Production Data." <u>Oil and
Gas Journal</u> 77, no. 28 (July 9, 1979):182.
    China's premier Hua, in his message to the National
People's Congress, revealed oil production in 1978 as 104 mil-
lion metric tons.

C615   "China's Biggest Chemical Fibre Plant." <u>Beijing Review</u> 22,
no. 28 (July 13, 1979):6-7.
    Reports that the Shanghai Petrochemical Works, China's
largest, went fully on line in June 1979.

C616   "China Slows to a Dragon's Pace." <u>Chemical Week</u> 125, no. 4
(July 25, 1979):37-38. illus.
    Analyzes the current state of the chemical industry in
China.

C617   "China's Oil: French Firm's Seismic Survey." <u>Business China</u>
(July 25, 1979):105.
    The French state-owned Elf-Aquitaine will join U.S. and
European firms to conduct seismic surveys of China's conti-
nental shelf.

C618   "China's Oil Goals Seen Realistic," by Lynn Brenner. Journal
        of Commerce (July 26, 1979):1, 29-30.
             Assesses China's chances for boosting her crude oil output
        by 10 times in the next 20 years.

C619   "Legs and Walking Stick--A Visit to the Qianjin Chemical
        Works," by Zhou Jin. Beijing Review 22, no. 30 (July 27,
        1979):13-17. illus.
             A profile of a major chemical plant, the bulk of whose
        facilities were imported from Japan. It is a division of the
        Yenshan (Yanshan) General Petrochemical Works (formerly the
        Peking General Petrochemical Works).

C620   "China's Offshore Oil Surveys." China Business Review 6,
        no. 4 (July/Aug. 1979):62. map.
             Provides an offshore map delineating major international
        oil firms' contracts and the locations of their seismic sur-
        veys of China's continental shelf from the Pohai (Bohai) Gulf
        to the South China Sea.

C621   "China Oil Production Growth Halted." Petroleum Economist
        46, no. 8 (Aug. 1979):337-38.
             Crude oil production in China might rise a mere 1.9% in
        1979 from the 1978 level, the lowest rate of growth in 20
        years.

C622   "Energy and Power," by George Lauriat. China Trade Report 17
        (Aug. 1979):6-7.
             Dissects China's new policy of seeking foreign technology
        and opening up her offshore area for exploration by foreign
        oil firms.

C623   "Ten U.S. Companies Join Search for Oil Off China," by Fox
        Butterfield. New York Times (Aug. 7, 1979):1, 27. map.
             At least ten major American oil firms will take part in
        comprehensive seismic surveys of China's continental shelf,
        particularly in the South China Sea.

C624   "Far East--The Focus Is on China." World Oil 189, no. 3
        (Aug. 15, 1979):225-28. illus., map.
             With the daily output of 2 million b/d, China emerged as
        the largest oil producer in the Far East in 1978. Her open-
        door policy toward the West has produced positive results,
        such as the signing of offshore exploration contracts with
        foreign oil firms and large purchases of the U.S. oil tech-
        nology and equipment.

C625   "In Pursuit of Oil," by Phijit Chong. <u>China Trade Report</u> 17
(Sept. 1979):3-4.
Discusses China's participation in the world oil trade and
prospects of her crude oil export to countries such as the
Philippines, Thailand, and Japan.

C626   "South China Sea Oil Search Mixes Economics, Politics," by
David Binder. <u>New York Times</u> (Sept. 2, 1979):E2. map.
Territorial disputes between China and her neighbors
Vietnam, the Philippines, and Taiwan highlight the importance
attached to undersea oil in the South China Sea.

C627   "China Trims Oil Production Goals." <u>Oil and Gas Journal</u> 77,
no. 36 (Sept. 3, 1979):36-38. illus.
China's targeted production of oil for 1979 shows only a
gain of 1.9% from that of 1978, a sharp drop from the previous
gain of 11% over 1977.

C628   "Pouring Trouble on Oily Waters," by George Lauriat and
Melinda Liu. <u>Far Eastern Economic Review</u> 105, no. 39
(Sept. 21, 1979):19-20. illus.
An account of the dispute between China and Vietnam over
territorial rights in the South China Sea in view of China's
contracts granted to the U.S. oil majors for conducting seismic surveys around China's Hainan Island and Vietnam's coastal
areas.

C629   "High Output Oil Field Developed in China." <u>Journal of Com-</u>
<u>merce</u> (Sept. 27, 1979):33.
News report on the development of a new oil field in central China in Honan (Henan) province, identified as the Nanyang
oil field.

C630   "China Calls in the Foreign Rigs," by Nayan Chanda. <u>Far</u>
<u>Eastern Economic Review</u> 106, no. 39 (Sept. 28, 1979):21.
Comments on the controversy surrounding China's decision
to invite foreign oil firms to conduct seismic surveys over
the sensitive Tonkin Gulf waters to which Vietnam àlso claims
territorial rights.

C631   "Mission Impossible for China: 400 Million Tons by 1990," by
V.G. Kulkarni. <u>Petroleum News</u> 10, no. 7 (Oct. 1979):8, 10.
Unless major onshore and offshore oil strikes are made,
there is no prospect for China to reach its target of 400 million tons by 1990, since crude oil output from China's onshore
fields seems to be declining.

C632   "Operating Practices in China's Dagang Oil Field," by J.R.
Pace. <u>World Oil</u> 189, no. 5 (Oct. 1979):93-97. illus. (some
color), map.
     The firsthand observations of a U.S. oil executive on the
Takang (Dagang) oil field during a recent trip to the site.
The author details all aspects of the operation, including
exploration, drilling, well completions, and offshore development.

C633   "Petroleum Developments in Far East in 1978," by G.L.
Fletcher. <u>Bulletin of the Association of American Petroleum
Geologists</u> 63 (Oct. 1979):1816-83. maps.
     Contains a detailed evaluation and survey of petroleum re-
sources of countries in the Far East, including China, and
covers petroleum geology, estimates on oil and gas reserves,
exploration, production, processing, and transportation for
the year 1978.

C634   "China Now Ranks No. 7 in World Power Generation--China '79,"
by Vaclav Smil. <u>Far Eastern Economic Review</u> 106, no. 40
(Oct. 5, 1979):81-82. illus.
     Presents the state of China's energy industry in 1979,
including its oil industry.

C635   "A Swell of Oil Disputes in the South China Sea." <u>Business
Week</u>, no. 2607 (Oct. 15, 1979):44.
     A capsule description of territorial disputes involving
China, Vietnam, Indonesia, and the Philippines over the South
China Sea.

C636   "Chinese Drilling Strike Oil." <u>Business China</u> (Oct. 17,
1979):142. map.
     China's oil drilling team struck oil at 6 out of 7 test
wells in the Gulf of Tonkin and the Pearl River estuary in
the South China Sea.

C637   "South China Sea Oil: Opportunities Amidst Political Uncertainties." <u>Business China</u> (Oct. 17, 1979):142. map.
     The likelihood of territorial disputes with Vietnam
dampens the otherwise bright prospects for new oil finds in
the Gulf of Tonkin.

C638   "China Claims Advanced Drilling Ability." <u>Oil and Gas Journal</u>
77, no. 42 (Oct. 22, 1979):24-25.
     China has achieved a relatively sophisticated level of
drilling technology as well as the capability to build onshore
and offshore drilling rigs.

C639   "New Oil Well in South China Sea." <u>Beijing Review</u> 22, no. 43
       (Oct. 26, 1979):3-4. illus.
           Reports on an oil find offshore in Kwangtung (Guangdong)
       province and the development of a new oil field at Nanyang in
       Honan (Henan) province in central China.

C640   "Energy and Power," by Vaclav Smil. <u>China Trade Report</u> 17
       (Nov. 1979):5-6. illus.
           Comments on Chairman Hua's report on China's energy situa-
       tion in 1979. The author detected a hint of imminent pros-
       pects for decline in China's crude oil output/reserve ratio
       if no new major oil field could be put on line.

C641   "Oil in the Tonkin Gulf," by Nayan Chanda. <u>China Trade Report</u>
       17 (Nov. 1979):17.
           China's decision to invite foreign oil firms to conduct
       seismic surveys off the Gulf of Tonkin could embroil her in a
       territorial dispute with Vietnam, since a section of blocks
       that China conceded was still a highly contested area between
       the two countries.

C642   "Search for Oil," by George Lauriat. <u>China Trade Report</u> 17
       (Nov. 1979):6-9. illus., map.
           China has given priority to developing her outer conti-
       nental shelf by permitting major American oil companies to
       take part in the seismic survey of offshore areas, including
       the sensitive zone in the Gulf of Tonkin.

C643   "India Considers China as Source of Crude," by Trevor
       Drieberg. <u>Journal of Commerce</u> (Nov. 6, 1979):32.
           Deals with the possibility of China's becoming a supplier
       of crude oil to India.

C644   "Chinese Chemicals Grow with Western Technology." <u>Chemical
       Marketing Reporter</u> 216 (Nov. 19, 1979):3, 61.
           The transfer of technology from the West has been largely
       instrumental in the rapid development of the chemical and
       petrochemical industries in China in the past few years.

C645   "Further Oil Strides by China Reported in Trade Magazine."
       <u>Wall Street Journal</u> (Nov. 26, 1979):23.
           Cites a report by <u>Oil and Gas Journal</u> (entry C646) on
       China's success in both onshore and offshore drilling in the
       South China Sea and Kwangtung (Guangdong) province.

C646   "Successes Spur Chinese Search Onshore and Off." <u>Oil and Gas
       Journal</u> 77, no. 47 (Nov. 26, 1979):19-21. illus., map.
           Describes China's redoubled effort in search of oil and
       gas in the South China Sea and Szechuan (Sichuan) province.

C647 "Sichuan Journal--Natural Gas," by Stephanie Green. China Business Review 6, no. 6 (Nov./Dec. 1979):14-16. illus.
    A detailed account of the visits to natural gas fields in Szechuan (Sichuan) province by a U.S. oil delegation sponsored by the National Council on U.S.-China Trade. The visit was highlighted by an inspection tour of Weiyuan gas field, which is located in the central part of the province.

C648 "Oil and Gas in China," by R.W. Scott. World Oil 189, no. 7 (Dec. 1979):55-61. illus. (some color), map.
    The author's 4th report on oil and and gas in China during his recent tour of the country. The latest article provides a vivid account of China's natural gas production practices and her success in ultra-deep drilling operations in search of natural gas.

C649 "China and Japan Sign Oil Development Pact." Journal of Commerce (Dec. 7, 1979):33.
    A news report on the accord reached between China and Japan on the joint exploration and development of oil and gas covering 10,000 square miles in Pohai (Bohai) Gulf. Japan will invest $500 million for a 42.5% interest in oil production.

C650 "The Lure of Chinese Oil," by Mohan Ram. Far Eastern Economic Review 106, no. 49 (Dec. 7, 1979):83.
    Reports on the offer by China to sell India up to one million tons of oil through a third party.

C651 "Chemical Industry in China on the March to Modernization," by Richard J. Seltzer. Chemistry and Engineering 57, no. 50 (Dec. 17, 1979):21-30. illus., map.
    A comprehensive survey of the state of China's chemical and petrochemical industries in 1979.

C652 "Chinese Petrochemicals Seen as Acquiring More Importance; Export Likely in Middle '80s." Chemical Marketing Reporter 216 (Dec. 17, 1979):7, 15. illus.
    Over 18 months, China has purchased numerous petrochemical plants from the West, which could result in a 400% increase in petrochemical production by 1985 and might herald China's emergence as a petrochemical exporter.

C653 "Oil in the Orient--Petroleum Men's Interest in Asia Gains; Stability Makes Up for Undramatic Finds," by Barry Newman. Wall Street Journal (Dec. 19, 1979):40. map.
    Analyzes the prospect and potential of finding new oil in the Far East, including China.

C654   "Sino Accent: Conservation, Exploration." <u>Oil and Gas Journal</u> 77, no. 53 (Dec. 31, 1979):42-43.
        China's energy plan for 1980 accents the acceleration of her offshore oil development and includes a 10% reduction of her oil consumption.

C655   "Soviets Doubt Accuracy of Chinese Oil Production Figures." <u>Oil and Gas Journal</u> 77, no. 53 (Dec. 31, 1979):42.
        The Soviet Union criticizes oil production data released by the Chinese government since 1970 as too exaggerated.

C656   "Offshore: The Petroleum Industry in the People's Republic of China, 1969-1978," by John B. Leach. <u>Chinese Economic Studies</u> 13, nos. 1-2 (Fall-Winter 1979-80):105-51.
        A fairly thorough evaluation of China's offshore oil development from 1969 to 1978, categorizing it in three phases. In phase one China stressed its indigenous efforts in exploiting undersea oil; in phase two, the expansion of oil logistics, particularly export-related facilities; in phase three, China will depend more on foreign technology and equipment to accelerate the development of her offshore oil reserve.

Chinese and Japanese Language Publications

(D)   Reference Works, Books, Documents, and Monographs

D1   Sensei-shō yuden chōsa shiryō陝西省油田調査資料[Research data on oil fields in Shensi province].   Dairen大連: Mantetsu, keizai chōsakai 満鉄経済調査会[Economic Research Association, South Manchurian Railway Co.], 19--.
A confidential investigation conducted by geologists F.G. Clapp et al. on the prospects for oil production and oil reserves, and the potential of oil fields at Yench'ang and other areas of northern Shensi province.   This report, obtained by Japanese authorities in Manchuria, was classified as secret.

D2   Taiwan yuden chōsa hōkoku台湾油田調査報告[A report on the investigation of Taiwan's oil fields].   Taipei台北: Taiwan sōtokufu minseibu shokusankyoku台湾総督府民政部殖産局[Bureau of Colonial Development, Civil Administration, Government General of Taiwan], 1910, 257 pp.   illus.
Probably one of the best-documented and earliest reports on the petroleum resources of Taiwan which grew out of a geological survey conducted by Japanese authorities after their occupation of the island.   It related the discovery of some oil and natural gas-bearing structures in the Chinsui 錦水 [Japanese: Kinsui] and Ch'uk'uangk'eng 出磺坑 [Japanese: Shutko-ko] areas of central Taiwan and pointed to the development of oil and gas fields in those areas.

D3   Bujun yubo ketsugan jigyō rengō kyōgikai kiroku 撫順油母頁岩事業組合協議会記録[Proceedings of the United Congress of Bujun (Fushun) Oil Shale Industries].   Dairen大連: Minami manshū tetsudō kabushiki kaisha 南満州鉄道株式会社[Southern Manchurian Railway Co.], 1925, 387 pp.   illus.
A rare document about developing the process of refining oil shale into shale oil in Fushun [Japanese: Bujun] county, Liaoning province, compiled by the Japanese interests in Manchuria.

D4 Bujun-san ketsugansekirō ni kansuru kenkyū
撫順産頁岩石爛に関する研究 [A Study of paraffin content in shale
oil produced at Bujun (Fushun)], by Yoshimi Konaka 小中義美].
Bujun?撫順: Bujun tankō kenkyūjo, 撫順炭礦研究所 [Bujun Colliery
Research Institute], 1935, 214 pp. illus.
  Report on the process of dewaxing paraffin from shale oil
produced at the Fushun [Japanese: Bujun] shale oil refinery.

D5 Chung-kuo ko sheng ran liao fen hsi 中國各省燃料分析 [The analysis of Chinese fuels], edited by Kai-ying Chin and Wu-chao
Hsia 金開英与夏武肇合撰. Nanking 南京: Nan-ching kuo-fu shih-yeh-
pu ti-chih tiao-ch'a-so 南京國府實業部地質調查所 [Geological Research
Institute, Ministry of Industry, Nanking Government], 1936,
106 pp. maps, charts.
  The examination and analysis of the supply of fuels such
as coal, oil, shale oil, and natural gas in the individual
provinces of China, with an appendix on ore and mineral analysis. In Chinese with an English summary at the end of the
book.

D6 Shina yuden chōsa shiryō 支那油田調查資料 [Research data on
China's oil fields]. 3 vols. Dairen 大連: Minami manshū
tetsudō kabushiki kaisha chōsabu 南満州鉄道株式会社調查部 [Research
Department, South Manchurian Railway Co.], 1937. illus. (some
color), maps.
  A confidential survey of China's petroleum geology and its
existing oil fields. The report was authorized and conducted
by the South Manchurian Railway Co. when controlled by Japanese
interests.

D7 Shōwa jūninendo Sansei yuboketsugan oyobi sekitanchi shisui
chishitsu chōsa hōkoku 昭和十二年度三姓油母頁岩及石炭地試錐地質調查報告
[Research report on the trial drilling of coal and oil shales
at Sansei (Sanhsing) in 1937]. Dairen 大連: Mantetsu san-
gyōbu 満鉄産業部 [Department of Industry, South Manchurian Railway Co.], 1937, 35 pp.
  A summary of Japanese geological prospecting for oil shale
deposits at Sanhsing [Japanese: Sansei] in Heilungkiang province, northern Manchuria. Stamped: top secret.

D8 Nekka-shō Kenshō-ken Gokashi fukin yuboketsugan chōsa hōkoku
熱河省建昌縣五家子附近油母頁岩調查報告 [Research report on oil shale
deposits in the vicinity of Nekka-shō Kenshō-ken Gokashi].
Dairen 大連: Mantetsu chōsabu 満鉄調查部 [Research Department,
South Manchurian Railway Co.], 1939, 44 leaves. illus.
  A geological survey of oil shale deposits in the mine at
Chiench'ang county, Jehol province [Japanese: Kenshō-ken
Nekka-shō] conducted to determine their commercial quantity.
The report was prepared under the auspices of the Japanese-
controlled railroad authorities.

D9   Kantō-shō Ōseiken Rashikō yuketsugan chōsa hōkoku
間島省汪清縣羅子溝油頁岩調査報告[Research report on oil shales from
Kantō-shō Ōseiken Rashikō]. Harbin 哈尔濱： Mantetsu hokuman
keizai chōsajo 満鉄北満経済調査所[North Manchuria Economic Research Institute, South Manchurian Railway Co.], 1939, 158 pp.
illus.
      Report on oil shale found at and shale oil refined in the
locality of Wangch'ing county in Chientao province [Japanese:
Ōsei-ken Kantō-shō], northern Manchuria.

D10 Bujun-san ketsuganyu no seisei to riyō ni kansuru kenkyū
撫順産頁岩油の精製と利用に関する研究[A study of the refining and
utilization of shale oils produced in Bujun (Fushun)], compiled by Bujun tankō kenkyūjo撫順炭礦研究所編[Bujun Colliery
Research Institute]. Dairen大連： Minami manshū tetsudō
kabushiki kaisha chōsabu南満州鉄道株式会社調査部[Research Department, South Manchurian Railway Co.], 1941, 330 pp. illus.
      A rare report on the operation of a shale oil refinery at
Fushun [Japanese: Bujun] by the Japanese prior to the end of
World War II.

D11 Chung-kuo k'uang-ch'an tzu-yüan 中國礦産資源[China's mineral
resources], compiled by Pin-fan Ch'en陳東範. Taipei 台北：
Chung-hua wen-hua ch'u pan shih-yeh wei-yüan-hui
中華文化出版事業委員會[Chinese Cultural Publication Commission],
1954, 200 pp. maps.
      This publication presents both a general survey and
evaluations of China's mineral resources up to 1952. Chapter
eight is devoted entirely to oil and gas resources. It provides a fairly broad examination of the geographic distribution of China's hydrocarbon resources and its reserves, and
the development, refining, and transportation of its petroleum
products. It also furnishes a breakdown of oil fields and
outputs of key provinces. Natural gas and oil shale have been
extensively evaluated and appraised in terms of their quality
and quantity, and particular attention has been paid to oil
shale and the location of shale oil deposits.

D12 Kung-fei shih-yu kung-yeh yen-chiu共匪石油工業研究[An inquiry
into the Chinese Communist oil industry]. Taipei 台北：
Chung-kuo kuo-min-tang chung-yang wei-yüan hui ti liu hsiao-
tzu中國國民党中央委員會第六小組[Section Six of the Central Committee, Chinese Nationalist Party], 1955, 35 pp.
      A critical evaluation of mainland China's petroleum
industry focusing on her supply and demand situation in the
1950s.

D13 Ti i ko shih-yu chi-ti--Yü-men 第一个石油基地－玉門[The first oil
base--Yümen], compiled by Lan Ho 何蘭編. Shanghai 上海：
Shang-hai mei-shu ch'u pan she 上海美術出版社 [Shanghai Fine Arts
Press], 1956, 40 pp.

A picturesque description of the growth and development of the Yümen oil field in Kansu province.

D14   T'ai-wan shih-yu ti-chih t'ao-lun-hui lun-wen chuan chi 台湾石油地质討論會論文專輯[Symposium on the petroleum geology of Taiwan], conducted by the Chinese Petroleum Corp. on 1 & 2 June 1956, in Taipei in celebration of the tenth anniversary of its establishment on June 1, 1946. Taipei 台北: Chung-kuo shih-yu ku-fen yu-hsien kung-szu 中國石油股份有限公司[Chinese Petroleum Corp.], 1957, 312 pp. illus. (67 leaves of plates).
      A collection of papers presented at the symposium on petroleum geology held in Taiwan in 1956. Texts either in Chinese or English.

D15   T'ai-wan shih-yu 台湾石油 [Petroleum in Taiwan], compiled by Ching-chi pu 經濟部編 [Ministry of Economic Affairs]. Taipei 台北: Ching-chi pu 該部 [Ministry of Economic Affairs], 1958, 12 pp. illus.
      A profile of the petroleum industry in Taiwan covering exploration, production, and refining.

D16   Chin-sui san-shih-pa hao ching tsuan-tso ch'eng-kung cheng-ming T'ai-wan shih-yu ch'u-ts'ang liang feng-fu. 錦水 三十八號井鑽鑿成功証明台湾石油儲藏量豐富 [The successful drilling of oil well no. 38 at Chinsui proves that Taiwan has rich petroleum reserves]. Taipei 台北: Hsing-cheng yüan hsin-wen chü 行政院新闻局 [Information Service, Executive Yüan], 1959, 24 pp.
      Describes the drilling of a wildcat well which struck substantial natural gas and some oil in the Chinsui area of central Taiwan.

D17   Chung-kuo k'uang-ch'an chih, 中國礦產志 [Minerals in China], edited by Hua-lung Wang 王華隆編著. Taipei 台北: T'ai-wan shang-wu yin shu-kuan 台湾商務印書館 [Taiwan Commercial Press], 1960, 184 pp. map.
      The publication covers the distribution and mining of China's major minerals and analyzes China's geological structure. Data on China's deposits of oil and gas and their production are based primarily on sources prior to 1949. However, the considerable space given to description of oil shale reserves in China includes a breakdown of the estimated holdings of nine provinces rich in oil shale deposits.

D18   Shih-yu ti-chih hsüeh lun wen-chi 石油地质学論文集第一集 [A collection of papers on petroleum geology], translated by Ch'eng-tu ti-chih hsüeh-yüan shih-yu chiao yen shih 成都地质学院石油研室譯 [Petroleum Teaching and Study Group, Ch'eng-tu Geological College]. Vol. 1. Peking 北京: K'o-hsüeh ch'u pan she 科學出版社 [Science Publishing Co.], 1961, 241 pp. illus., chart.

Translations of foreign papers on tectonics done by teachers and students at Ch'engtu Geological College.

D19　Sekiyu sangyō ni miru Chūgoku no jiryoku kōsei 石油産業にみる中国の目力更生[China's self-reliance as seen in her petroleum industry], by Nobuo Shinoda 篠田信男. Tokyo 東京: Asahi shimbun chōsa kenkyūshitsu 朝日新聞調査研究室[Office of Investigation and Research, Asahi Shimbun], 1966, 59 pp. map.
　　This report attempts an analysis of China's effort to develop her petroleum industry through self-reliance and with minimal foreign assistance in the fifties and sixties. It includes supplements describing China's oil fields and refineries.

D20　T'ien-jan-ch'i ch'ien-tsai shih-ch'ang hsü-yao tiao-ch'a pao-kao 天然気潜在市場需要調査報告[The potential market demand for natural gas: an investigative report], edited by Chung-kuo sheng-ch'an-li chi mao-i chung-hsin中國生產力及貿易中心[Chinese Productivity and Trade Center]. 2 vols. Taipei 台北: Chung-kuo sheng-ch'an-li chi mao-i chung-hsin Chung-kuo shih-yu ku-fen yu-hsien kung-szu T'ai-wan kuan-yeh 中國生產力及貿易中心中國石油股份有限公司台湾営業　[Chinese Productivity and Trade Center, Chinese Petroleum Corp., Taiwan Administration], 1967?
　　This survey of the western region of Taiwan indicates that a sizable potential market exists for the consumption of natural gas by its residential and industrial communities.

D21　Ta-lu shih-yu kung-yeh hsien-shih 大陸石油工業現勢[The current condition of the petroleum industry in mainland China], edited by Ching-chi pu 經濟部編 [Ministry of Economic Affairs]. Taipei 台北: Ching-chi pu 該部 [Ministry of Economic Affairs], 1968, 121 pp. illus.
　　A general survey and evaluation of mainland China's petroleum industry, examining her oil reserves and assessing her exploration, production, refining, consumption, and transportation of petroleum.

D22　Chung-kuo-ti shih-yu tsu-yüan chi ch'i k'ai-fa 中國的石油資源及其開發 [Petroleum resources and their development in China], by Cheng-siang Chen 陳正祥. Hong Kong 香港: Hsiang-kang chung-wen ta-hsüeh yen-chiu-yüan ti-li yen-chiu chung-hsin香港中文大学研究院地理研究中心[Geographical Research Centre, Graduate School, Chinese University of Hong Kong], 1968, 17 pp. map.
　　A survey and assessment of China's oil and natural gas reserves, including oil shale and production, which empha-sizes China's efforts to develop new oil resources since 1950. In Chinese with a summary in English. A revised edition under the title Chung-kuo-ti shih-yu kung-yeh 中國的石油工業 [The Petroleum Industry in China], 34 pp., was

issued by the same publisher as a research report (no. 52)
in 1972.

D23   Chung-kung nien pao 中共年報 [Yearbook on Chinese Communism].
      Taipei 台北： Chung-kung yen-chiu tsa-chih she 中共研究雜誌社
      [Institute for the Study of Chinese Communist Problems],
      1969--. charts, tables.
          One of the most exhaustive and informative annual studies
      of Chinese Communist affairs reporting on almost every facet
      of their political, economic, social, cultural, military, and
      other activities for that year. This yearbook contains an
      annual survey of mineral resources (including natural gas
      and oil shale) in China which encompasses a wide spectrum of
      reserves of natural petroleum and covers the distribution of
      oil shale and of major basins with oil-bearing potential.
      Most of the information published here is drawn from foreign
      or domestic intelligence sources.

D24   Erh-shih nien lai chih Chung-kuo shih-yu 二十年來之中國石油
      [Chinese petroleum in the last twenty-five years].
      Taipei 台北： Chung-kuo shih-yu ku-fen yu-hsien kung-szu
      中國石油股份有限公司 [Chinese Petroleum Corp.], 1971, 242 pp.
      illus.
          An historical description of the Chinese Petroleum
      Corporation from its inception at the end of World War II.
      Gives an account of the expansion and development of refining
      and the petrochemical industry in Taiwan.

D25   Shih-yu jen shih hua 石油人史話 [Discourses on the history of
      the Chinese Petroleum Corporation by its executives], by
      Hung-hsün Ling et al. 凌鴻勛等人著. Taipei 台北： privately
      published, 1971, 596 pp. illus.
          This publication was issued in commemoration of the
      twenty-fifth anniversary of the Chinese Petroleum Corporation
      from its inception in 1947 in mainland China. The memoirs,
      narrated by top executives of Taiwan's oil company, record
      the continuous growth and development of this state-owned oil
      concern from the time of its establishment in China to its
      reemergence in Taiwan. It also chronicles all the major
      events of those twenty-five years in the history of Taiwan's
      oil industry.

D26   T'ai-wan shih-yu chi t'ien-jan-ch'i chih t'an-k'an yü k'ai-fa
      台湾石油及天然气之探勘与開發[The exploration for and development
      of petroleum and natural gas in Taiwan]. Miaoli, Taiwan
      苗栗台湾： Chung-kuo shih-yu ku-fen yu-hsien kung-szu T'ai-wan
      yu-k'uang t'an-k'an ch'u 中國石油股份有限公司台湾油礦探勘處[Taiwan
      Oil Field Exploration Office, Chinese Petroleum Corp.], 1971,
      121 pp. illus.
          A historical description of the exploration for and pros-
      pecting of oil and gas in Taiwan emphsizing progress made
      since the end of World War II.

D27   T'ai-wan shih-yu hua-hsüeh kung-yeh chih chien-chieh
台湾石油化學工業之简介 [A summary of the petrochemical industry
in Taiwan], edited by Chung-kuo shih-yu ku-fen yu-hsien kung-
szu Kao-hsiung lien-yu-ch'ang 中國石油股份有限公司高雄煉油廠編
[Kaohsiung Refinery, Chinese Petroleum Corp.]. Kaohsiung,
Taiwan 高雄台湾: Chung-kuo shih-yu ku-fen yu-hsien kung-szu
Kao-hsiung lien-yu-ch'ang 該廠 [Kaohsiung Refinery, Chinese
Petroleum Corp.], 1971, 23 pp. illus.
      A brief sketch of the petrochemical industry's development
in Taiwan.

D28   T'ai-wan shih-yu t'an-k'an chi-yao 台湾石油探勘紀要 [A summary of
petroleum exploration in Taiwan]. Miaoli, Taiwan 苗栗台湾:
Chung-kuo shih-yu ku-fen yu-hsien kung-szu T'ai-wan yu-k'uang
t'an-k'an ch'u 中國石油股份有限公司台湾油礦探勘處 [Taiwan Oil Field
Exploration Office, Chinese Petroleum Corp.], 1971, 237 pp.
illus.
      A historical description of the prospecting and develop-
ment of petroleum resources in Taiwan, with emphasis on
progress in the post-war period.

D29   Shih-yu t'an-k'an 石油探勘 [Petroleum exploration], by Ta-ch'ing
yu-t'ien (shih-yu t'an-k'an) pien hsieh tsu 大庆油田石油探勘編寫組
[Compiling and Editing Group (Oil Exploration), Tach'ing Oil
Field]. Shanghai 上海: Shang-hai jen-min ch'u pan she
上海人民出版社 [Shanghai People's Publishing Co.], 1972, 96 pp.
illus.
      This booklet describes simply those theories and tech-
niques of petroleum geology relating to oil prospecting which
have been applied in the development of Chinese petroleum
resources.

D30   P'ou-shih kung-fei shih-yu lien-chih chi-shu yü ch'an-p'in
p'in-chih 剖視共匪石油煉製技術与產品品質 [The analysis of petroleum
refining technology and products quality in China]. Taipei
台北: Ching-chi pu 經濟部 [Ministry of Economic Affairs],
1974, 226 pp. illus.
      Analyzes and assesses the petroleum-processing industry
and the quality of petroleum products in mainland China.
Although Communist China has been absorbing expertise in oil
technology from Soviet Russia and Rumania since 1950 and has
developed her refining industry to a certain level, the
quality of her oil refining technology and of her petroleum
products is only equivalent to that of the Western nations
in the 1950s and remains far behind the advanced level of
those countries in the 1970s.

D31   Ta-lu shih-yu tzu-yüan chih fen-pu 大陸石油資源之分佈 [The dis-
      tribution of petroleum resources in mainland China], edited
      by Ching-chi pu 經濟部 [Ministry of Economic Affairs]. Taipei
      台北: Ching-chi pu 該部 [Ministry of Economic Affairs], 1974,
      33 pp. illus., map.
          This booklet describes all aspects of Chinese petroleum
      resources including oil, natural gas, and oil shale.

D32   Ta-ch'ing 大庆 [Tach'ing], compiled by Ta-ch'ing ke-ming wei-
      yüan-hui 大庆革命委員會編 [Tach'ing Revolutionary Committee].
      Shanghai 上海: Shang-hai jen-min ch'u pan she 上海人民出版社
      [Shanghai People's Publishing Co.], 1974, 138 pp. all illus.
      (some color).
          A collection of 101 photos, 34 of which are in color.
      These describe men and women at work in Tach'ing oil base and
      emphasize its thriving atmosphere and the integration of
      agriculture and industry there.

D33   Yüeh chin chung ti yu-t'ien chien-she 跃進中的油田建設 [Oil field
      construction is advancing], edited by Shang-hai jen-min ch'u
      pan she 上海人民出版社編輯 [Shanghai People's Publishing Co.].
      Shanghai 上海: Shang-hai jen-min ch'u pan she 該社 [Shanghai
      People's Publishing Co.], 1975, 48 pp. illus.
          A collection of articles on the building of China's oil
      fields.

D34   Ta-lu fei-ch'ü shih-yu hua-hsüeh kung-yeh 大陸匪区石油化學工業
      [Petrochemical industry in mainland China], by Ching-chi pu
      經濟部 [Ministry of Economic Affairs]. Taipei 台北: Ching-chi
      pu 該部 [Ministry of Economic Affairs], 1975, 28 pp.
          One of the most thorough analyses and evaluations of the
      state of China's chemical and petrochemical industries com-
      piled by the authority in Taiwan.

D35   Ta-kang yu-t'ien chih kai-shu 大港油田之概述 [A synopsis of
      Takang oil field], compiled by Ta-lu shih-yu chi hua-hsüeh
      kung-yeh yen-chiu tsu Chung-kuo shih-yu ku-fen yu-hsien
      kung-szu 大陸石油及化學工業研究組中國石油股份有限公司 [Study Group for
      Petroleum and the Petrochemical Industry in mainland China,
      Chinese Petroleum Corp.]. Taipei 台北: Chung-kuo shih-yu
      ku-fen yu-hsien kung-szu 中國石油股份有限公司 [Chinese Petroleum
      Corp.], 1975, 73 pp. illus., maps.
          This is probably the most comprehensive examination and
      analysis of China's major onshore and offshore oil fields yet
      published outside China. The historical development of
      Takang oil field's onshore and offshore operations and pros-
      pects have been analyzed and evaluated.

D36   Sen-kyūhyaku-shichijūyonen no Chūgoku keizai no
        ugoki, sekiyu kanren sangyō o chūshin ni shite:
        Chūgoku keizai kankei chōsa hōkoku sho 1974年の中国経済の
        うごき：石油関連産業を中心にして：中国経済関係調査報告書
        [The trend of China's economy in 1974, as focused on
        the petroleum-related industries: a research report
        on the Chinese economy]. Tokyo 東京： Nichū keizai
        kyōkai 日中経済協会 [Sino-Japanese Economic Association],
        1975, 160 pp.
            This investigative report examines and assesses the
        performance of China's economy in 1974 vis-a-vis the
        progress achieved in her petroleum and petrochemical
        industries. The report considers the role played by
        China's oil-related industries as vital to the overall
        improvement of the Chinese economy in the year 1974.

D37   Chūgoku sekiyu: sono genjō to kanōsei 中国石油その現状と可能性
        [China's petroleum: its current state and possibilities],
        by Susumu Yabuki 矢吹晋編. Tokyo 東京： Ryukei shosha 竜溪書舎
        [Ryukei Publishing Co.], 1976, 215 pp. illus.
            A comprehensive examination and survey of China's
        petroleum industry--including exploration, reserves,
        production, refining, and transportation. Her production
        and export potential in the future are emphasized, with
        an eye on Japan as a market for oil exports.

D38   Gendai no Chūgoku keizai: sekiyu to shakaishugi kensetsu
        現代の中国経済：石油と社会主義建設 [Contemporary Chinese economics:
        petroleum and the building of socialism], by Kazuo Yamanouchi
        山内一男. Tokyo 東京： Chuō kōron sha 中央公論社 , 1976, 185 pp.
            Discusses the priority given to the petroleum industry
        in China's overall economic planning and building of a
        socialist state.

D39   Ch'ien chin a: Chin-shan kung-ch'eng 前进啊金山工程 [March
        forward: the Chinshan project], compiled by Shang-hai
        jen-min ch'u pan she 上海人民出版社 [Shanghai People's Publish-
        ing Co.]. Shanghai 上海： Shang-hai jen-min ch'u pan she 該社
        [Shanghai People's Publishing Co.], 1976, 135 pp. illus.
            Describes the planning and building of Shanghai's biggest
        industrial project, the vast Shanghai General Petrochemical
        Work located in Chinshan county on the outskirts of Shanghai.
        The first stage of construction involved eighteen processing
        units to be completed in 1976. About 50,000 workers took
        part in the initial phase of building in January 1974.

D40   Ta-lu fei-ch'ü shih-yu shu ch'u hsi-t'ung chih yen-chiu
        大陸匪区石油輸儲系統之研究 [A study of mainland China's petroleum
        transportation and storage system], edited by Ching-chi pu
        經濟部編 [Ministry of Economic Affairs]. Taipei 台北： Ching-
        chi pu 經濟部 [Ministry of Economic Affairs], 1976, 42 pp.
        illus., maps.

One of the most comprehensive studies of China's petroleum transportation capabilities as seen in her extensive oil pipe-lines, oil terminals, oil port facilities, tanker fleets, etc. Compiled by the Taiwan authority.

D41   Kao-chü Mao chu-hsi-ti wei-ta ch'i chih tsou wo kuo tzu-chi kung-yeh fa-chan-ti tao-lu: Ta-ch'ing yu-t'ien tang-wei shu-chi ke-wei-hui chu-jen Sung Chen-ming tsai ch'üan-kuo kung-yeh Ta-ch'ing hui-i shang-ti fa-yen高举毛主席的伟大旗帜走我国自己工业发展的道路大庆油田党委书记革委会主任宋振明在全国工业大庆会议上的发言 [Raising the great banner of Chairman Mao and walking our own road to industrial development: a speech given by Sung Chen-ming, Director of the Revolutionary Committee and Party Secre-tary at Tach'ing oil field, at the Conference on State Industry held at Tach'ing], by Chen-ming Sung 宋振明. Peking 北京: Jen-min ch'u pan she 人民出版社[People's Publishing Co.], 1977, 46 pp.

An important speech on the historical development of Tach'ing oil field given by the Director of the Revolutionary Committee of the oil field, Sung Chen-ming (currently Minister of the Petroleum Industry), at the National Congress for Learning from Tach'ing held at Tach'ing oil field and in Peking in 1977. The address emphasizes Chinese self-reliance in the projected development of the oil field as a sprawling petroleum and petrochemical complex over a sixteen-year period.

D42   Chūgoku no sangyōbetsu kensetsu seika: denryoku sangyō, un'yu-kanren sangyō, tetsukō kōgyō, sekiyu kōgyō 中国産業別建設成果:電力産業運輸関連産業鉄鋼工業石油工業[China's nation-building industrial achievements: the transportation, steel, and petroleum industries]. Tokyo 東京: Nichū keizai kyokai 日中経済協会 [Sino-Japanese Economic Association], 1978, 173 pp. illus.

A comprehensive assessment of some key industries in China including the petroleum industry.

D43   Ta-lu shih-yu hua-hsüeh kung-yeh 大陸石油化学工業 [Mainland China's petrochemical industry], edited by Ching-chi pu 經濟部 [Ministry of Economic Affairs]. Taipei 台北: Ching-chi pu 該部 [Ministry of Economic Affairs], 1978, 40 pp. illus.

A comprehensive review of China's petrochemical industry. This is an updated and revised version of the same title pub-lished by the ministry in 1975 (entry D34).

D44   Chung-kuo-ti shih-yu 中国的石油 [Petroleum resources in China], by Cheng-siang Chen 陳正祥. Hong Kong 香港: T'ien-ti t'u-shu kung-szu 天地圖書公司[T'ien-ti Book Co.], 1979, 299 pp. illus., map, tables.

The most up-to-date, informative, and well-balanced evaluation and survey of China's petroleum industry and trade in the Chinese language. Drawing on original Chinese sources as well as Western studies, the author describes and analyzes various facets of China's oil industry ranging from its historical development, oil-bearing basins and oil fields, offshore prospects, shale oil, and natural gas to exploration, production, refining, transportation, and export potential. The work is lavishly illustrated with eighteen plates of photos. The bibliography in Chinese, Japanese, and English is also extensive.

D45   Ta-ch'ing yu-t'ien chi-shu ke-hsin tzu-liao hsüan 大庆油田技术革新资料选 [Selected materials on technological innovations at Tach'ing oil field], edited by Ta-ch'ing chi ko tzu-liao san chieh ho pien hsieh tsu 大庆技革资料三结合编写组 [Tach'ing Technological Innovation Triple Editorial Group]. Shanghai 上海: Jen-min ch'u pan she 人民出版社 [People's Publishing Co.], 1979-. illus.
    A collection of articles, essays, and treatises on technical and technological innovations in the recent development and operation of the Tach'ing (Daqing) oil field.

D46   Chung-kuo shih-yu kung-yeh fa-chan shih 中国石油工业发展史 [The history of the development of China's petroleum industry], edited by Li-sheng Shen 申力生编. 3 vols. Peking 北京: Shih-yu kung-yeh ch'u pan she 石油工业出版社 [Petroleum Industry Publishing Co.], 1980--. illus.
    This publication provides the most complete and well-documented coverage of the development of petroleum and natural gas resources in ancient and contemporary China. The already-published first volume (122 pp.) covers the period from 250 B.C. to 1840 A.D. while the more recent and contemporary epochs will be dealt with in the second and third volumes scheduled to be published in 1981 and 1982. The first volume introduces systematically Chinese scientific and technological achievements in geology and oil technology as demonstrated in their early success in the discovery, exploration, and exploitation of petroleum and natural gas resources. In addition, the volume also chronicles in detail the development of Chinese technology in oil and gas well drilling, techniques of extraction, geological studies, and utilization of petroleum and natural gas, the last illustrated through the existence of China's first natural gas well in Szechuan (Sichuan) province in the eighteenth century.

(E) Articles

E1 "Chūgoku no sekiyu jijō" 中国の石油事情[Petroleum situation in China]. Kaigai shigen jōhō 海外資源情報[Report on Overseas Natural Resources] 7 (Jan. 1973):1-34.
    A comprehensive review and evaluation of the petroleum industry in China.

E2 "Chūgoku tairiku no sekiyu--sono zenyō to taigai senryaku (tokushū)"中国大陸の石油ーその全容と対外战畧(特集)[Petroleum in mainland China--its full story and foreign strategies (special issue)]. Shūkan daiamondo 週刊ダイアモソド[Diamond Weekly] 61 (Oct. 27, 1973):24-40. illus.
    A comprehensive survey of China's petroleum industry, its prospects, and its oil trade policy.

E3 "Chūgoku no sekiyu kōgyō no genjō to tembō-- Wagakuni to no kyōdō kaihatsu no kanōsei (jo ge)"中国の 石油工業の現状と展望ー我が国との共同開発の可能性（上下） [China's petroleum industry, its current situation and prospects-- the prospects of joint exploration with Japan (pts. 1-2)]. Jūkagaku kōgyō tsūshin 重化学工業通信 [Heavy Chemical Industry Newsletter] pt. 1, 2855 (Jan. 11, 1974):1-4; pt. 2, 2856 (Jan. 18, 1974):1-13.
    A brief survey of the petroleum industry in China with a discussion of possible Sino-Japanese development of Chinese oil.

E4 "Sekiyu taikoku o mezasu Chūgoku no kaihatsu senryaku" 石油大国を目ずす中国の開発戦略[China's developmental strategy aims for leadership in oil production]. Tōyō keizai 東洋経済 [Far Eastern Economics] 3788 (Feb. 26, 1974):66-69. illus.
    Examines and evaluates the current production of crude oil in China and predicts its future potential.

E5 "Waga michi o yuku Chūgoku sekiyu kaihatsu"我が道を行く中国石油開発 [China travels her own road in oil development], by Hideo Ōno 大野英男. Chūgoku keizai kenkyū geppō中国経済研究月報[Chinese Economic Research Monthly] 99 (March 1974):1-58.
    An extensive examination and analysis of China's oil industry emphasizing self-reliance and independence.

E6 "Chūgoku--sekiyu hinkoku kara no dasshutsu--Nichū enerugi kyō-ryoku taisei no tembō"中国-石油貧国からの脱出ー日中エネルギー協力体制の展望 [China's exit from oil scarcity--the prospect of Sino-Japanese cooperation on energy], by Kazuo Asakai 浅洲一男. Gendai Chūgoku 現代中国 [Contemporary China] 9 (Spring 1974):4-16. illus.
    Analyzes the development of China's petroleum industry and the possibility of Japanese participation in the exploitation of China's oil deposits.

E7   "Chūgoku sekiyu no maizō to seisan"中国石油の埋蔵と生産[Chinese petroleum reserves and production]. Shōwa Dojin 昭和同人 7 (July 1974):19-27.
      On China's offshore and onshore oil deposits, her current crude oil production and future trends.

E8   "Chūgoku saidai no yuden--Taikei"中国最大の油田ー大慶[China's largest oilfield--Tach'ing], by Hiroshi Kaneshige 金重紙. Sekai shūhō 世界週報 [Weekly World Report] (July 30, 1974): 40-43.
      An eyewitness report on life at the Tach'ing oil field written by a Japanese reporter. It describes the Tach'ing rural-industrial complex of 400,000 workers and their dependents.

E9   "Yung Mao chu-hsi che-hsüeh ssu-hsiang chih-tao chao yu chi" 用毛主席哲學思想指导找油气[Chairman Mao's philosophical thought guides us in search of oil and gas], by Chung-kuo kung-ch'an-tang Ta-kang yu-t'ien wei-yüan-hui中国共産党大港油田委員会 [Takang Oil Field Committee, Chinese Communist Party]. Hung ch'i 紅旗 [Red Flag] 276 (Aug. 1974):75-81.
      An account of the exploration and development of the Takang oil field through the application of Mao's doctrine.

E10   "Chūgoku no sekiyu kōgyō (tokushū)"中国の石油工業（特集） [Petroleum industry in China (special issue)]. Ajia keizai jumpō アジア経済旬報 [Asian Economic Bulletin] 943 (Aug. 1-10, 1974):3-30. tables.
      A general survey of the state of China's petroleum industry and its prospects.

E11   "Chūgoku no genyū kakaku to kyokyū no tembō" 中国の原油価格と供給の展望[China's crude oil price and supply prospects], by Hideo Ōno大野英男. Chūgoku keizai kenkyū geppō 中国経済研究月報 [Chinese Economic Research Monthly] 105 (Sept. 1974):1-34.
      An optimistic estimate of China's oil production and export capabilities for the future.

E12   "Saikin ni okeru Chūgoku keizai no ugoki-- sekiyu kanren sangyō no hatten o chūshin ni"最近にお ける中国経済の動きー石油関連産業の発展を中心に　　[Recent trends in China's economy--focusing on the development of petroleum-related industries], by Kazuo Yamanouchi山内一男. Kaigai shijō 海外市場 [Overseas Markets] 24 (Sept. 1974):20-27.
      Discusses the leading role played by petroleum and petroleum-related industries in the development of China's economy.

E13 "Kono me de mita Taikei yuden--Chūgoku kokudo kaihatsu no
ichi moderu"この目でみた大慶油田-中国国土開発の一モデル[An eyewitness
report on the Tach'ing oil field--China's developmental
model], by Ryūzō Yamashita 山下竜三. Ekonomisuto エコノミスト
[Economist] (Sept. 3, 1974):52-55.
　　The profile of the Tach'ing oil field as reported by a
Japanese who made an inspection tour of the oil facilities
at Tach'ing.

E14 "Hakusha kakaru Chūgoku no sekiyu kaihatsu"拍車かかる中国の石油開発
[China's oil development in high gear], by Katsuhiko Hama
浜勝彦. Ekonomisuto エコノミスト[Economist] 1 (Oct. 1974):
30-34.
　　An appraisal of China's oil production and an estimate
of its future potential.

E15 "Chung-kuo ta-lu jan-liao kou-ch'eng pien-hua-ti t'an-t'ao"
中国大陸燃料構成変化的探討[Examination of changes in fuel composition in mainland China], by Yeh-hui Hsiao 蕭野暉. Chung-kung
yen-chiu 中共研究 [Study of Chinese Communism] 10 (Oct. 1974):
47-60. tables.
　　Despite substantial gains in China's oil production, coal
continues to constitute a major share of her energy consumption.

E16 "Taikei yuden no mezasu mono--kōshin kara senshin e no migoto
na tenshin"大慶油田のめざすもの一後進から先進へのみごとな転身 [The aim
of the Tach'ing oil field--an excellent transformation from
the underdeveloped to the advanced], by Ryūzō Yamashita
山下竜三. Ajia keizai jumpo アジア経済旬報 [Asian Economic
Bulletin] 949 (Oct. 1-10, 1974):6-18.
　　Describes the achievements and development of the Tach'ing
oil field.

E17 "Shogaikoku kara mita Chūgoku no sekiyu jijō"
諸外国から見た中国の石油事情[China's oil situation as viewed
by various foreign nations]. Chōsa geppō (Ōkurasho)
調査月報(大蔵省) [Research Monthly (Ministry of Finance)]
12 (Dec. 1974):40-62.
　　Compares Japanese and Western views of the petroleum
industry in China.

E18 "Chūgoku no sekiyu sangyō" 中国の石油産業 [China's petroleum
industry]. Sekiyu kaihatsu jihō 石油開発事報 [Petroleum
Development Times] 24 (Dec. 1974):15-41. tables.
　　A comprehensive survey of China's oil industry--its
reserves, exploration, production, refining, transportation,
and oil exports.

E19　"Taikō yuden no kembun" 大港油田の見聞 [Observations from the Takang oil field]. Ajia keizai jumpō アジア経済旬報 [Asian Economic Bulletin] 24 (Dec. 1974):15-41.

　　　　An extensive report and description of Takang oil field written by a Japanese reporter who recently visited there.

E20　"Chung-kuo shih shih-yu tsu-yüan tsui feng-fu-ti kuo-chia" 中國是石油資源最豐富的國家 [China is the nation richest in petroleum resources], by Ping-ti Ho 何炳棣. Ch'i-shih nien-tai 七十年代 [Seventies] 61 (Feb. 1975):6-14. illus.

　　　　A fairly comprehensive survey of the petroleum industry in China. The author estimates China's oil reserves and production and, on the basis of his readings and the impressions made during his visit to China, paints a very optimistic picture of oil production in the 1970s and 1980s. His projection of China's crude oil production, under present circumstances, appears to be unrealistic.

E21　"Sanyukoku Chūgoku no shōrai" 産油国中国の将来 [China's future as an oil producer]. Ekonomisuto エコノミスト [Economist] 12 (Feb. 1975):46-50. illus.

　　　　Discusses China's current oil production and her potential as an oil producer.

E22　"Tui Chung-kung shih-yu sheng-ch'an shu-ch'u chi yün-shu neng-li-ti yen hsi" 対中共石油生産輸出及運輸能力的研析 [An examination of Chinese Communist oil production, exports, and transportation capabilities], by Heng Ch'ien 錢亨. Chung-kung yen-chiu 中共研究 [Study of Chinese Communism] 4 (Apr. 1975):98-103. tables.

　　　　Despite an expansion in her crude oil output, China will not be able to export substantial quantities of oil due to such limited transport facilities as pipelines, oil port facilities, etc.

E23　"Chūgoku no sekiyu shigen to taigai bōeki" 中国の石油資源と対外貿易 [China's oil resources and her foreign trade]. Asahi Ajia ribyū 朝日アジアリビュー [Asahi Asia Review] 3 (Sept. 1975):48-55. illus.

　　　　An evaluation of China's oil reserves and the prospects for oil export to Japan.

E24　"Fukyō to Chūgoku to sekiyu sangyō (Shinshun kandan)" 不況と中国と石油産業（新春散談）[Economic slump, China, and the petroleum industry (New Year's discussion)], by Yoshitaro Wakimura and Keinosuke Idemitsu 脇村義太郎出光計助. Sekiyu bunka 石油文化 [Petroleum Culture] 1 (Jan. 1976):6-31. illus.

　　　　Discusses Japan's economic stagnation as it affects the import of large quantities of Chinese crude oil by the Japanese oil refining industry.

E25   "Chūgoku enerugī baransu no mitōshi"中国：エネルギーバランスの見通し
[China: energy balance projections]. Chūgoku keizai kenkyū
geppō 中国経済研究月報 [Chinese Economic Research Monthly] 122
(Feb. 1976):71-108.
      The Japanese translation of the CIA's research report on
China's energy situation entitled China: Energy Balance
Projections (entry B22).

E26   "Chūgoku ni okeru saikin no sekiyu jijō"中国における最近の石油事情
[Recent developments concerning China's oil]. Chōsa geppō
(nai-chō) 調査月報（内調）[Research Monthly (Domestic Research)]
5 (May 1976):1-17.
      A general survey of the current state and future prospects
for China's petroleum industry.

E27   "Chūgoku no sekiyu kagaku kōgyō"中国の石油化学工業 [Petrochemical
industry in China], by Yoshihiko Hirakawa平川芳彦. Kagaku
keizai化学経済 [Chemistry Economics] 6 (June 1976):70-76.
illus.
      Discusses the development and current state of China's
petrochemical industry.

E28   "'Jiryoku kōsei' no riron--Taikei yuden no shisō kōzō (Chūgoku
ripōto)"自力更生の理論－大慶油田の思想構造(中国リポート)[The theory of
"self-reliance"--ideological framework of Tach'ing oil field
(report on China)], by Masanori Kikuchi菊地昌典. Gekkan
ekonomisuto月刊エコノミスト [Monthly Economist] 8 (Aug. 1976):
100-108. illus.
      Discusses China's self-help doctrine, based on her
previous experience, as motivating her not to seek foreign
assistance in building her oil industry.

E29   "Taiwan no sekiyu kagaku kōgyō"台湾の石油化学工業 [Petrochemical
industry in Taiwan]. Mitsubishi Shoji kaigai joho
三菱商事海外情報 [Mitsubishi Trading Company Overseas Report]
9 (Sept. 1976):22-28.
      A general survey of the current state and future prospects
of Taiwan's petrochemical industry.

E30   "Iwayuru 'Chūgoku sekiyu' no ţembō" 所謂中国石油の展望 [The prospects for what is called "China's oil"], by Kōkichi Miyata
宮田幸吉. Seikei ronsō (Kokushikan dai)政経論叢（国士館大）
[Treatises on Politics and Economics (Kokushikan University)]
25 (Nov. 1976):51-76.
      A comprehensive analysis of China's petroleum reserves,
exploration, and production with an eye to its future export
potential.

E31　"Taikei no akahata" 大慶の赤旗 [Tach'ing's red banner], by Kaoru Ishida 石田かおる.　Asahi Ajia ribyū 朝日アジアリビュー[Asahi Asia Review] 4 (Dec. 1976):92-97.　illus.
　　Describes China's success in the development of this vast oil-producing base primarily through her own efforts.

E32　"Chung-kuo shih-yu kung-yeh kai-k'uang" 中國石油工業概況[Synopsis of the petroleum industry in China], by Chung-te Huang 黄種德. Ch'i-shih nien-tai 七十年代 [Seventies] 88 (May 1977):7-10.
　　A realistic appraisal of China's oil industry--its reserves, production, refining, transportation, exports, and future potential.

E33　"Ta-lu shih-yu kung-yeh chih fa-chan ch'ing-k'uang" 大陸石油工業之發展情況 [The condition of mainland China's petroleum industry], by Chu-yüan Cheng 鄭竹園. Ming pao yüeh-k'an 明報月刊 [Ming Pao Monthly] 5 (May 1977):2-7.　illus., map.
　　This is a summary and updating of the author's earlier publication, China's Petroleum Industry: Output Growth and Export Potentials, published in 1976 (entry B32).　The article examines the historical development of China's oil industry, the growth of crude oil production, the geographic distribution of China's oil fields, and China's refining capabilities and transportation facilities.　An objective and informative presentation of the state of China's oil industry in a succinct form.

E34　"Ta-lu shih-yu-ti shu-ch'u ch'ien-li" 大陸石油的輸出潛力 [China's petroleum export potential], by Chu-yüan Cheng 鄭竹園.　Ming pao yüeh-k'an 明報月刊 [Ming Pao Monthly] 6 (June 1977):37-40. tables.
　　A condensed and updated version of the author's earlier publication, China's Petroleum Industry: Output Growth and Export Potentials, published in 1976 (entry B32).　The article primarily dissects China's oil production in the light of her domestic consumption and thereby estimates possible surplus for export.　It stresses the need for vast capital investments in the oil industry if China wishes to export substantial quantities of crude oil in the 1980s.

E35　"Chūgoku sekiyu sangyō no genjō to tembō"中国石油産業の現状と展望 [China's petroleum industry: its present situation and prospects], by Chieh-yün T'ien 田潔雲.　Jimmin Chūgoku人民中国 [People's China] 296 (Feb. 1978):12-20.　illus. (some color).
　　A general introduction to China's developing petroleum industry illustrated with numerous colorful photographs.

E36   "Tsai lun Chung-kuo-ti shih-yu wen-t'i" 再論中国的石油问题
       [Another discussion of China's petroleum problems], by Ping-ti
       Ho 何炳棣. Ch'i-shih nien-tai 七十年代 [Seventies] 97 (Feb.
       1978):68-77.
          The updated and revised version of the author's mono-
       graph published in the same journal in May 1975 (entry E20).
       Assesses oil reserves and production in China. Chinese
       geologist Li Szu-kuang's theory about China's oil is analyzed.
       Also examines key features of oil exploration in China and
       refutes some Western geologists' bias toward China's oil
       prospects. Compared with his previous paper, the author makes
       a somewhat cooler and more sober assessment of China's future
       trends in production and export of oil. This is a well-
       balanced and more factual presentation of China's overall oil
       situation in 1977.

E37   "Genyu no tai-Nichi yushutsu gonengo sen-gohyaku man ton"
       原油の对日輸出五年後千五百万吨[Crude oil exports to Japan will
       reach fifteen million tons in five years], by Toshio Maeda
       前田寿夫. Sekai shūhō 世界週報 [Weekly World Report] (Feb. 21,
       1978):16-20. illus.
          A realistic projection of oil imports from China based on
       the expanded trade made possible by the conclusion of a
       twenty-billion dollar trade accord between China and Japan
       in 1978.

E38   "Hsin-hsin-hsiang-jung-ti Ta-ch'ing yu-t'ien" 欣欣向榮的大庆油田
       [The prospering Tach'ing oil field], by Ta-ch'ing yu-t'ien
       pao-tao tsu 大庆油田報导组[Tach'ing Oil Field Information
       Group]. Ti-li chih-shih 地理知識 [Geographic Knowledge] 4
       (Apr. 1978):4-6. illus.
          A historical description of the Tach'ing oil field from
       its beginnings in 1959 through two decades of growth and
       development. The article emphasizes the success of this field
       and its contribution to the building of a petroleum industry
       in China.

E39   "Hsi-k'an Chung-Jih shih-yu kuan-hsi"細看中日石油關係[A close
       look at the Sino-Japanese oil relationship], by Ta-ching Liu
       劉大經. Ch'i-shih nien-tai 七十年代 [Seventies] 101 (June
       1978):17-21.
          The author reviews China's oil exports in recent years
       and focuses on her potential oil exports to Japan in light of
       the February 1978 Sino-Japanese trade agreement calling for a
       twenty-billion dollar exchange of goods and services over the
       next eight years. Liu discusses the mutual benefits to be
       derived from this trade pact. He foresees the possibility of
       trade exceeding the dollar amount set by this agreement in
       view of China's decision to develop ten more large oil fields
       in the 1980s, each capable of producing twenty million tons
       of crude oil.

E40　"Chung-kuo yüan yü hsi-fang ho-tso k'ai-fa shih-yu chin-shu
k'uang-ts'ang"中国願与西方合作開發石油金屬礦藏 [China is willing to
develop her oil, metal, and mineral resources jointly with
the West]. Ta-kung-pao (Overseas edition) 大公報 (Sept. 1,
1978):1.

　　According to the Kyōdō News Service, Kang Shien, Deputy
Premier of China, has said that China is planning joint ventures with foreign oil concerns to develop oil resources on
her outer continental shelf in the Pohai Gulf, the East China
Sea, the Yellow Sea, and the South China Sea.

E41　"Chung-kuo tseng chien hsin lien-yu-ch'ang" 中国増建新煉油廠
[China has built many new oil refineries]. Ta-kung-pao
(Overseas edition) 大公報 (Sept. 2, 1978):1.

　　To meet an enormous increase in the production of crude
oil, China has raised her oil-refining capacity by almost
400 percent during the past ten years. About 57 percent of
this gain is attributed to improvement of existing refining
facilities, while the remainder results from the building of
new oil refineries.

E42　"Sheng-li yu-t'ien tseng-ch'an hsün-su" 勝利油田増産迅速[Produc-
tion at the Shengli oil field rises rapidly]. Ta-kung-pao
(Overseas edition) 大公報 (Sept. 8, 1978):1.

　　According to a Kyōdō News Service report from Chinan,
Shantung province, crude oil production at the Shengli oil
field has increased so rapidly that it will match that of
Tach'ing oil field within the next three to five years. The
Japanese estimated oil output at Shengli at twenty million
tons in 1977.

E43　"Ta-ch'ing jeng k'o k'ai-ts'ai san-shih nien"大慶仍可開採三十年
[Tach'ing oil field can still last another thirty years], by
Chung Ming 鐘明. Ta-kung-pao (Overseas edition) 大公報
(Sept. 11, 1978):4.

　　According to Professor Peng Chi-jui of the University of
Hong Kong, geologists at the Tach'ing oil field have begun to
explore for oil at depths greater than 1,500 meters to ascer-
tain which zones in the strata are oil-bearing. Thus far the
oil-bearing zone above 1,500 meters is already adequate to
last for more than thirty years.

E44　"Po-hai-wan yu-t'ien nien ch'an-liang mu-piao wei san-ch'ien-
wan kung tun"渤海灣油田年産量目標為三千萬公噸[Target of annual oil
production at Pohai Gulf oil fields set at thirty million
tons]. Ta-kung-pao (Overseas edition) 大公報 (Sept. 15, 1978):
1.

　　According to reports released by Kyōdō, a Japanese news
agency, China is currently developing offshore oil fields in
the Pohai Gulf. She expects that they will be producing 30-50
million tons of crude oil in the near future. A high Chinese

official is quoted as saying that China has succeeded in
drilling a discovery well which produces 1,000 tons of crude
oil a day, and that about seventy wells are currently being
developed in the same area.

E45   "Po-hai Chu-chiang ts'ai yu chi-hua Chung-Jih ta-ch'eng ho-tso
hsieh-i"渤海珠江採油計劃中日達成合作協議[Sino-Japanese accord on
planning oil exploration off Pohai and Pearl River]. Ta-kung-
pao (Overseas edition) 大公報 (Sept. 25, 1978):2.
       According to a Reuters news report, China and Japan have
reached an agreement on the joint development of offshore oil
in the Pohai Gulf of north China and the Pearl River delta off
Kwangtung province in south China. Informed sources say that
the possibility of participation by American firms in the
Pearl River project was not ruled out.

E46   "Lu-Ning shu-yu-kuan-tao chien ch'eng"魯寧輸油管道建成[Building
of the Shantung-Nanking oil pipeline completed]. Ta-kung-pao
(Overseas edition) 大公報 (Sept. 28, 1978):1.
       China's first north-to-south oil trunk line has been
completed. The building of this large-caliber pipeline per-
mits crude oil produced at the Shengli and Huapei oil fields
to be shipped directly to the oil terminal at the port of
Nanking. This one-thousand-kilometer-long pipeline should
help improve the distribution of fuel supplies to central and
southern China.

E47   "Shōri yuden--daini no Taikei"勝利油田－第二の大慶[Shengli oil
field--the second Tach'ing],by Akio Sugino杉野明夫. Ajia
keizai jumpō アジア経済旬報 [Asian Economic Bulletin] 1093
(Oct. 1, 1978):5-14.
       The author describes his impressions of the Shengli oil
field during his visit in July 1978. He presents a broad
profile of the oil base with its oil wells dotting the farm-
land of Shantung province. Sugino estimates the crude oil
output at Shengli to have been in the vicinity of eleven
million tons in 1974.

E48   "Jen-ch'iu yu-t'ien chien ch'eng" 任丘油田建成[Building of the
Jench'iu oil field completed]. Ta-kung-pao (Overseas edition)
大公報 (Oct. 4, 1978):1.
       A high-yielding oil field (with oil wells producing an
average of over 1,000 tons a day) has gone on line in Hopeh
province. The oil field was built with such speed that less
than two years passed between initial drilling and completion
of the field. The Jench'iu oil field is currently one of the
most productive in China.

E49   "Ch'ai-ta-mu fa-hsien ta yu-t'ien" 柴達木發現大油田 [A large oil
field discovered in Tsaidam basin]. Ta-kung-pao (Overseas
edition) 大公報 (Oct. 6, 1978):1.

    According to <u>Kuang-ming jih-pao</u>, a daily newspaper pub-
lished in Peking, a large oil field has been discovered in
the Tsaidam basin. From the east to the west side of the
basin oil and natural gas have been found. Tsaidam basin is
located in the northwestern part of Tsinghai province. It
has a total area of more than two hundred thousand square
kilometers, larger than all of Great Britain.

E50   "Ho-pei Ts'ang-chou ch'eng shih-yu ch'eng" 河北滄州成石油城
[Ts'angchou in Hopeh province has become an oil city]. Ta-
kung-pao (Overseas edition) 大公報 (Oct. 31, 1978):1.

    The city of Ts'angchou in Hopeh province is fast becoming
a petrochemical town largely due to its proximity to the oil
field at Jench'iu. The city has a petrochemical facility
which can process 750,000 tons of crude oil supplied by the
Jench'iu oil field. In addition, the state has also built a
fertilizer plant with an annual capacity of 1,000,000 tons.
Most of the facilities at the plant have been imported from
abroad.

E51   "Chung-kuo k'ai-fa Po-hai yu-t'ien Jih-pen t'i-kung chi-shu
ho-tso" 中國開發渤海油田日本提供技術合作 [Japan offers technological
cooperation for the development of China's Pohai oil fields].
Ta-kung-pao (Overseas edition) 大公報 (Nov. 3, 1978):1.

    Informed sources in Tokyo report that China and Japan are
expected to reach an accord next month on a $2 billion loan
to develop oil fields in the Pohai Gulf during the next eight
to ten years. The Japanese will be responsible primarily for
the exploration of undersea oil in the gulf and, in return,
will receive crude oil from China. The quantity of oil has
not yet been determined.

E52   "Te o musubu Chūgoku to meijā no omowaku"
チを結ぶ中国とメイジャーの思惑 [China's connection with oil
majors--a puzzle], by Nobuo Miyajima 宮島信夫. Ekonomisuto
エコノミスト [Economist] 45 (Nov. 11, 1978):16-21. illus.

    The Japanese seem to be alarmed by close contacts between
China and the international oil majors. Since China's ten-
year plan (which started in 1978) calls for the rapid development of her offshore oil (an undertaking which requires the
investment of huge sums of foreign capital), the international
oil companies seem poised and ready to take the opportunity
to cooperate with China. Should this connection bear fruit,
it is bound to have a deep impact on Japan's trade strategy.

E53   "Chung-kuo shih-yu yüeh yang ch'u kuo ming-nien k'ai-shih shu wang Nan Mei"中國石油越洋出國明年開始輸往南美[Beginning next year Chinese petroleum is to be shipped to South America]. Mei-chou hua-ch'iao jih-pao 美洲華僑日報[American Overseas Chinese Daily (New York)] (Nov. 11, 1978):1.

     An accord has been reached in Peking between China and Brazil on the exchange of China's crude oil for Brazil's iron ore and steel products. Beginning in 1979, China will ship 1 million tons of crude oil to Brazil and in 1980, 1.5 million tons. From 1981 on the quantity of Chinese crude will be increased each year.

E54   "An-hui An-ch'ing shih-yu hua tsung ch'ang t'ou ch'an" 安徽安慶石油化總厰投產 [Anhui's Anch'ing General Petrochemical Works begin production]. Ta-kung-pao (Overseas edition) 大公報 (Dec. 30, 1978):1.

     After more than three years of construction the Anch'ing General Petrochemical Works began production on Dec. 4, 1978. This marked the completion of the first stage of construction for the petrochemical complex, which includes a refinery, a thermal power plant, a fertilizer plant, and a repair shop. The fertilizer plant is one of thirteen imported from abroad and has a production capacity of 300,000 tons of ammonium sulphate and 520,000 tons of urea.

Appendix: Resource Libraries

Generally speaking, many large business and science libraries throughout the U.S. and Canada contain good-sized collections of books and periodicals concerning the petroleum industry in China. In addition, numerous East Asian libraries provide rich resources and a wide variety of information on China, including her petroleum industry. The following is a list of special libraries whose collections reflect the availability of literature cited in this bibliography.

BUSINESS AND SCIENCE LIBRARIES.

Arkansas

Murphy Oil Corp. Library. Jefferson Ave., El Dorado, Ark. 71730.

California

Atlantic Richfield Corp. Headquarters Library. 515 Flower St.,
    Los Angeles, Calif. 90051.

Kern County Library System. 1315 Truxtun Ave., Bakersfield, Calif.
    93301.

Long Beach Public Library. 101 Pacific Ave., Long Beach, Calif.
    90802.

Los Angeles Public Library, Science & Technology Department.
    630 W. Fifth St., Los Angeles, Calif. 90071.

Standard Oil of California Library. 225 Bush St., San Francisco,
    Calif. 94120.

Stanford University, School of Earth Sciences, Branner Earth Sciences
    Library. Stanford, Calif. 94305.

Union Oil of California Library. Brea, Calif. 92621.

Appendix

Colorado

Colorado School of Mines, Arthur Lakes Library. Golden, Colo. 80401.

District of Columbia

U.S. Department of Energy. Energy Library. 20 Massachusetts Ave. N.W., Washington, D.C. 20545.

Library of Congress. Science and Technology Division. 10 First St. S.E., Washington, D.C. 20540.

Illinois

Standard Oil of Indiana, Central Research Library. Naperville, Ill. 60540.

University of Illinois (Urbana/Champaign), Geology Library. 223 Natural History Building, Urbana, Ill. 61801.

Indiana

Indiana State University Library. 8600 University Blvd., Evansville, Ind. 47712.

Kansas

University of Kansas, Earth Sciences Library. 318 Lindley Hall, Lawrence, Kans. 66044.

Massachusetts

Harvard University, Geological Sciences Library. Cambridge, Mass. 02138.

Harvard University Library, Graduate School of Business Administration, Baker Library. Boston, Mass. 02163.

Stone & Webster Engineering Corp., Technical Information Center, Library. 245 Summer St., Boston, Mass. 02107.

New Jersey

Exxon Research and Engineering Co., Research Information Center (RIC). Linden, N.J. 07036.

New Mexico

Roswell Public Library. 301 N. Pennsylvania Ave., Roswell, N.Mex. 88201.

Appendix

New York

Engineering Societies Library. 345 E. 47th St., New York, N.Y.
    10017.

Mobil Corp., Secretariat Library. 150 E. 42 St., New York, N.Y.
    10017.

New York Public Library, Science and Technology Division. 42nd St.
    and 5th Ave., New York, N.Y. 10018.

Oklahoma

AMOCO Production Co. Research Center, Library. 4502 E. 41 St.,
    Tulsa, Okla. 74102.

Tulsa City-County Library, Business and Technology Department.
    400 Civic Center, Tulsa, Okla. 74103.

University of Oklahoma, Geology Library. 830 Van Vleet Oval,
    Rm. 102, Norman, Okla. 73019.

Pennsylvania

Carnegie Library of Pittsburgh. 4400 Forbes Ave., Pittsburgh, Pa.
    15213.

Franklin Institute Library. 20th & the Parkway, Philadelphia, Pa.
    19103.

Pennsylvania Geological Society Library. 916 Executive House,
    2nd & Chestnut Sts., Harrisburg, Pa. 17120.

Pennsylvania State University, Earth and Mineral Sciences Library.
    105 Deike Building, University Park, Pa. 16802.

Texas

Exxon Corp., U.S.A. General Service Library. Houston, Tex. 77001.

Midland County Public Library, Science-Technology Section. Midland,
    Tex. 79701.

University of Texas, Austin, General Libraries, Geology Library.
    Austin, Tex. 78712.

Canada

AMOCO Canada Petroleum Co., Library. Calgary, Alta. T2P0Y2.

University of Manitoba, Elizabeth Dafore Library. Manitoba R3T 2N2.

Appendix

EAST ASIAN LIBRARIES

California

Claremont Colleges, Nonnold Library. Claremont, Calif. 91711.

Hoover Institution on War, Revolution, and Peace, East Asian Collection. Stanford University, Stanford, Calif. 94305.

University of California, Berkeley, East Asiatic Library. Durant Hall, Berkeley, Calif. 94720.

University of California, Los Angeles, Oriental Library. 405 Hilgard Ave., Los Angeles, Calif. 90024.

Connecticut

Trinity College, Moore Collection of Far East. 300 Summit St., Hartford, Conn. 06106.

Yale University Library, East Asian Collection. 120 High St., New Haven, Conn. 06520.

District of Columbia

Howard University, Founder's Library, Bernard B. Hall Collection (Southeastern Asian Collection). Washington, D.C. 20059.

Library of Congress, Asian Division. Washington, D.C. 20540.

Hawaii

University of Hawaii, Library, Asia Collection. The Mall, Honolulu, Hawaii 96822.

Kansas

University of Kansas, Watson Library, East Asian Library. Lawrence, Kans. 66045.

Maryland

University of Maryland, Library, East Asia Collection. College Park, Md. 20742.

Massachusetts

Harvard University Library, Harvard-Yenching Library. Cambridge, Mass. 02138.

Appendix

Michigan

Michigan State University Libraries, International Library, East Asia
    Collection. East Lansing, Mich. 48824.

University of Michigan Libraries, Asia Library. Ann Arbor, Mich.
    48109.

New Jersey

Princeton University, Gest Oriental Library and East Asian Collec-
    tions. 317 Palmer Hall, Princeton, N.J. 08540.

New York

Cornell University, John M. Olin Library, Wason Collection of China
    & the Chinese. Ithaca, N.Y. 14853.

Columbia University Libraries, East Asian Library. 300 Kent Hall,
    New York, N.Y. 10027.

New York Public Library, Oriental Division. 5th Ave. & 42nd St.,
    New York, N.Y. 10018.

U.S. Military Academy Library. West Point, N.Y. 10996.

Pennsylvania

University of Pittsburgh, East Asian Library. 234 Hillman Library,
    Pittsburgh, Pa. 15260.

Rhode Island

Brown University, John Hay Library, East Asia Collection.
    20 Prospect St., Providence, R.I. 02912.

Texas

University of Texas, Austin, General Libraries, Asian Collection.
    Austin, Tex. 78712.

Washington

University of Washington Libraries, East Asia Library. Seattle,
    Wash. 98195.

Author Index

Adie, W.A.C., C310
Alley, Rewi, C248
Armentrout, F., C449
Asakai, Kazuo, E6
Aulderridge, Larry, C350, C491
Awanohara, Susumu, C288, C511

Backman, W.A., C500, C574, C579
Bakke, Donald R., C201, C307
Bartke, Wolfgang, B43
Baxter, Frank I., C64
Binder, David, C626
Bosum, W., C94
Bourne, Eric, C495
Brenner, Lynn, C613, C618
Brittain, Mary Z., C119
Brown, Wilfred, C458
Bujun tankō. kenkyujo [Bujun
    (Fushun) Colliery Research
    Institute], D10
Burchett, Wilfred, C164, C179
Burt, Richard, C470
Butterfield, Fox, C301, C454,
    C623

Cardew, John, C14
Carlson, Sevinc, B50
Central Intelligence Agency. See
    U.S. Central Intelligence
    Agency
Chan, Leslie, B20
Chanda, Nayan, C630, C641
Chang, Chün, C190, C342
Chang, Chun, C205
Chang, Keng, B7
Chang, Kuei-sheng, C31

Chang, Kwang-shih, C216, C368
Chang, Raymond, A31
Chang, Stanley, C107
Chen, Cheng-siang, D22, D44
Chen, H.H., C159
Chen, Ju-chin, C141
Chen, Ke-chung, C38
Chen, King C., C337
Ch'en, Pin-fan, D11
Chen, Si Z., C589
Chen, W.K., C89
Cheng, Chang-lu, C460
Cheng, Chu-yuan, B32, C315, C333,
    C435, E33, E34
Ch'eng-tu ti-chih hsüeh-yüan
    shih-yu chiao yen shih
    [Petroleum Teaching and Study
    Group, Ch'eng-tu Geological
    College], D18
Chi, Wen, C309
Chiang, Shan-hao, B48, C369,
    C371, C372, C373
Chien, Hsu, C17
Chien, Heng, E22
Chien, Suyen, C118
Chin, Chi, C272
Chin, Kai-ying, D5
Chin, Yun-shan, C93
China Energy Resources Study Team
    (Taiwan), B13
China, People's Republic of.
    Chung-kuo kung-ch'an-tang
    Ta-kang yu-t'ien wei-yüan-hui
    [Takang Oil Field Committee,
    Chinese Communist Party], E9

171

China, Republic of (Taiwan).
  Ching-chi pu [Ministry of
  Economic Affairs], D15, D21,
  D30, D31, D34, D40, D43
China, Republic of (Taiwan).
  Chung-kuo kuo-min-tang
  chung-yang wei-yüan-hui ti
  liu hsiao tzu [Section Six
  of the Central Committee,
  Chinese Nationalist Party],
  D12
China, Republic of (Taiwan).
  Hsing-cheng yüan hsin-wen chü
  [Information Service, Execu-
  tive Yüan], D16
China, Republic of (Taiwan).
  Industrial Development and
  Investment Center, Ministry
  of Economic Affairs, B18.
  See also China, Republic of
  (Taiwan), Ching-chi pu
Chinese Petroleum Corp. (Taiwan),
  C97. See also Chung-kuo
  shih-yu ku-fen yu hsien kung
  szu
Chong, Phijit, C364, C625
Chou, J.T., C63, C142
Chu, Chi-lin, C18
Chu, Godwin, C19
Chung, C.T., C29
Chung, Hsin, C171
Chung, Wen, C235
Chung-kuo sheng-ch'an-li chi
  mao-i chung-hsin [Chinese
  Productivity and Trade
  Center], D20
Chung-kuo shih-yu ku-fen yu hsien
  kung-szu [Chinese Petroleum
  Corp.], D14, D24, D35. See
  also Chinese Petroleum Corp.
  (Taiwan)
Chung-kuo shih-yu ku-fen yu hsien
  kung-szu Kao-hsiung lien-yu-
  ch'ang [Kaohsiung Refinery,
  Chinese Petroleum Corp.], D27
Chung-kuo shih-yu ku-fen yu hsien
  kung-szu T'ai-wan yu-k'uang
  t'an-k'an ch'u [Taiwan Oil
  Field Exploration Office,
  Chinese Petroleum Corp.],
  D26, D28

Chuter, A., C609
Clapp, Frederick G., C2
Cohen, Jerome A., C279, C351
Colm, Peter W., B35
Connel, Horton, C204
Cook, James, C430
Copper, John Franklin, B34
Cranfield, J., C266
Crook, David, C255
Cullison, A.E., C473, C483, C496,
  C568, C569

Dean, Genevieve C., C188
Dorsey, J.J., C156
Drieberg, Trevor, C643

Emery, K.O., C82, C100
Esposito, Bruce J., B41, C380

Fan, P.F., C438
Fan, Paul H., C265, C303
Fan, Shih-ching, C93
Fang, Po-hsi, C489
Farnsworth, Clyde H., C604
Fingar, Thomas, B30
Fletcher, G.L., C633
Foreign Languages Press (Peking),
  B17
Freeman, John, C440
Frost & Sullivan, B49
Fuller, Myron L., C1, C2

Gardner, Frank J., C126, C136,
  C211
Garth, Bryant G., C226
Gavin, Martin J., C8
Geiger, Robert E., C241
Ghosh, Sailen, C30, C37
Giles, William E., C260
Gill, Ranjit, C442
Glenn, W., C212
Goodstadt, Leo, C148, C221
Green, Stephanie R., C423, C544,
  C647
Gupta, Dipankar, C424
Gupta, Harmala Kaur, C424

Hama, Katsuhiko, E14
Hao, Paul L.C., C55
Hardy, Randall W., B33, B46

Subject Index

Albania, assistance to, C479
Alternate fuels, B50, B53, C304, C404, C507. See also Oil shales
Anch'ing (Anqing) General Petrochemical Works, E54

Badger, Ltd., C195, C433
Bethlehem Singapore (Bethlehem Steel Co.), Ltd., C447, C608
Bohai No. 1. See Pohai No. 1
British Petroleum Co., C582, C586, C609
Bujun. See Oil shales
Bush, George, C436

Changlin Oil Refinery, C418
Chekiang (Zhejiang) Oil Refinery, C543
Chem-Oil Industries, C592
Chemical industry. See Petrochemical industry
China-Korea friendship oil pipeline, C302
Chinese Petroleum Corp., A8, C458. See also Taiwan
Chinhuangtao (Qinhuangdao) tanker terminal, C262, C339. See also Oil tanker terminals
Coastal Corporation. See Coastal State Gas Co.
Coastal State Gas Co., C540, C541, C542, C551, C559, C601
Continental-Emsco Co., C567
Continental Oil Co. (CONOCO), C202

Continental shelf. See Offshore oil development, Crude oil reserves
Crude oil, export of, B31, B32, B33, B34, B35, B36, B39, B41, B43, B46, B47, B50, B52, B53, C158, C178, C180, C184, C200, C221, C225, C237, C238, C258, C266, C278, C280, C289, C297, C300, C305, C308, C315, C319, C334, C338, C340, C351, C355, C363, C380, C391, C393, C440, C495, C548, C591, C602, C625, E11, E22, E23, E34. See also Oil policy
-to Asia, C181, C338, C625
-to Australia, C383
-to Brazil, C548, C549, E53
-to Hong Kong, C187, C506, C550
-to India, C643, C650
-to Italy, C537, C548
-to Japan, B41, C71, C180, C187, C221, C238, C284, C285, C300, C305, C312, C313, C318, C364, C401, C430, C437, C483, C505, C509, C515, C536, C568, C625, E11, E37, E39
--Quality problem (Paraffin content), C288, C312, C340, C355, C364, C504, C530
-to Philippines, C300, C355, C441, C625
-to Thailand, C187, C355, C625
-to U.S., C482, C484, C540, C541, C542, C548, C551, C559, C577, C591, C601, E52

Title Index

"Active Faulting and Tectonics in China," C389
"Adventures of the 'Gang of Seventeen' in China," C585
"Airomagnetic Survey of Offshore Taiwan," C94
"An Appraisal of the Changing Fuel Structure on the China Mainland,"
      C213
"An-hui An-ch'ing shih-yu hua tsung-ch'ang t'ou ch'an" [Anhui's
      Anch'ing General Petrochemical Works begin production], E54
"Another Super-Deep Well," C499
"Asian Alliances and Chinese Oil," C439
"Assessing China's Oil Industry," C341
"At Taching: Red Banner on Industrial Front," C120
"Aussie Trial Purchase of Chinese Crude Oil Raises Hopes for Sales,"
      C383
"Awaiting the Rush of Chinese Crude, If Any--China '77," C405

"Basic Chemical and Petrochemical Industry in the Republic of
      China," C89
"Behind the Bamboo Curtain: An All-out Drive to Boost Output," C20
"Bethlehem Delivers Jack-Up to China," C608
Bibliography of Petroleum Geology of Mainland China, A27
"Big Advances in the Oil Industry," C95
"Big Hopes for China's Energy Resources," C220
"Big Increase in Crude Oil," C131
"Black Gold," C456
"Black Gold and the Red Flag--Taching Impressions (III)," C372
"Black Magic," C67
"BP Search in First Phase to Yellow Sea Oil Auction," C609
"British Petroleum to Conduct a Survey in China's Yellow Sea," C582
"Builders of the Taching Oil Field," C324
"Building Oil Industry through Self-Reliance--A Visit to the New Oil
      Pipeline as well as Taching and Takang Oilfields (I)," C227
"Building Oil Industry through Self-Reliance--A Visit to the New Oil
      Pipeline as well as Taching and Takang Oilfields (II)," C228
"Built with Soviet Aid--The Lanchow Oil Refinery," C23
Bujun-san ketsugan-yu no seisei to riyō ni kansuru kenkyū [A study
      of the refining and utilization of shale oils produced in Bujun
      (Fushun)], D10

Bujun-san ketsugansekirō ni kansuru kenkyū [A study of paraffin con-
tent in shale oil produced at Bujun (Fushun)], D4
Bujun yubo ketsugan jigyō rengō kyogikai kiroku [Proceedings of the
United Congress of Bujun (Fushun) Oil Shale Industries], D3

"Canton Gears Up for Oil," C285
"Cartel's New Competitor," C278
"Chairman Hua inspects Taching," C374
"Chai-ta-mu fa-hsien ta yu-t'ien" [A large oil field discovered in
Tsaidam], E49
"Chemical Industry in China on the March to Modernization," C651
Ch'ien chin a: Chin-shan kung-ch'eng [March forward: the Chinshan
project], D39
China (by Choon-ho Park), B21
"China" (in Energy and U.S. Foreign Policy), C175
"China: A Bid for the Future," C303
"China: A Bid for U.S. Help in Unlocking Its Oil," C145
"China Agrees to Buy Seven Drilling Rigs from Unit of LTV," C557
"China Agrees to Sell 634,900 Barrels of Oil to Refinery in Italy,"
C537
"China: A Nation Reaches for the Energy Age," C129
"China and Its Oil," C362
"China and Japan Sign Oil Development Pact," C649
"China and Offshore Oil: The Tiao-yu Tai Dispute" (in China's
Changing Role in the World Economy), C222
"China and Offshore Oil: The Tiao-yu Tai Dispute" (in Stanford
Journal of International Studies), C239
China and United States Policy, B45
"China Asks Petrocanada and Ranger Oil to Participate in Offshore
Exploration," C523
"China Awards Pullman Kellogg Pact for Petrochemical Facility," C560
"China Beckons the West--Watching the World," C491
"China Boasts Record Drilling, Production, New Oil Line," C532
"China Building a Sizable Oil Industry," C267
"China Builds Own Rigs for Pohai Search," C328
"China Buys 7 Oil Rigs," C558
"China Buys Two Drilling Rigs," C400
"China Calls in the Foreign Rigs," C630
"China Claims Advanced Drilling Ability," C638
"China Completes Major Oilfield in Jenchiu," C524
"China Completes Pipeline to Step Up Oil Exports," C230
"China: Development of the Oil Industry," C165
"China Discovers Offshore Oil Field," C453
"China Driving to Boost Oil Production," C327
"China Emerging as an Oil Power," C301
China: Energy Balance Projections, B22
"China Estimates Offshore Potential at 10-Billion Tons of Crude,"
C525
"China Expanding Purchases of Oil Equipment, Technology," C578
"China Expands Oil Search," C385
"China Expects Oil-Exploration Pacts," C604

China's Offshore Oil: Application of a Framework for Evaluating Oil
     and Gas Potentials under Uncertainty, B24
"China's Offshore Oil Surveys," C620
"China's Offshore Petroleum," C421
"China's Oil," C180
"China's Oil and Gas Reserves and Resources: A Review," C261
"China's Oil Export to Asia," C181
"China's Oil Extraction Techniques May Soon Be Exported to Japan,"
     C446
"China's Oilfield," C357
"China's Oil Fields Hold Promise of Foreign Participation," C477
"China's Oilfields Train Technical Personnel," C588
"China's Oil: French Firm's Seismic Survey," C617
China's Oil Future: A Case of Modest Expectations, B46
"China's Oil Goals Seen Realistic," C618
"China's Oil Hurt by Revolution," C75
China's Oil Industry: A Background Survey, B25
"China's Oil Industry Develops Rapidly," C154
"China's Oil Is Offered to America," C484
"China's Oil: Oil Under Disputed Waters," C276
"China's Oil Policy" (in Oil Weekly), C11
"China's Oil Policy" (in Yale Review), C337
"China's Oil Policy" (in Post-Mao China and U.S.-China Trade), C351
China's Oil: Problems and Prospects, B50
"China's Oil Production and Consumption," C182
"China's Oil Prospects," C380
"China's Oil Trade in the 1980s: A Closer Look," C338
"China's Oil Workers Gird Up for New Task," C489
"China's Perplexing Energy Triangle," C602
"China's Petroleum Industrial Front Scores Substantial Achievements,"
     C161
China's Petroleum Industry, B31
"China's Petroleum Industry" (in Military Review, Nov. 1969), C87
"China's Petroleum Industry" (in Indian Quarterly, Oct./Dec. 1975),
     C289
"China's Petroleum Industry--(I)" (in Far Eastern Economic Review,
     Sept. 1965), C42
"China's Petroleum Industry--(II)" (in Far Eastern Economic Review,
     Oct. 1965), C43
"China's Petroleum Industry Advances with Big Strides on Road of
     Self-reliance," C85
"China's Petroleum Industry--An Enigma," C204
"China's Petroleum Industry Continues Fast Advance in 1974," C233
"China's Petroleum Industry Develops with Greater, Faster, Better,
     and More Economical Results," C102
"China's Petroleum Industry Is Booming and Overfulfills Its 1973
     Production Quotas," C198
China's Petroleum Industry: Output Growth and Export Potentials,
     B32
"China's Petroleum Industry Strides Ahead," C290

Chung-kuo k'uang ch'an chih [Minerals in China], D17
Chung-kuo k'uang-ch'ang tzu-yüan [China's mineral resources], D11
"Chung-kuo shih shih-yu tsu-yüan tsui feng-fu-ti kuo-chia" [China is the nation richest in petroleum resources], E20
Chung-kuo shih-yu kung-yeh fa-chan shih [The history of the development of China's petroleum industry], D46
"Chung-kuo shih-yu kung-yeh kai-k'uang" [Synopsis of the petroleum industry in China], E32
"Chung-kuo shih-yu yüeh yang ch'u kuo ming-nien k'ai-shih shu wang Nan Mei" [Beginning next year, Chinese petroleum to be shipped to South America], E53
"Chung-kuo ta-lu jan-liao kou-ch'eng pien-hua-ti t'an-t'ao" [Examination of changes in fuel composition in mainland China], E15
Chung-kuo-ti shih-yu [Petroleum resources in China], D44
Chung-kuo-ti shih-yu tsu-yüan chi ch'i k'ai-fa [Petroleum resources and their development in China], D22
"Chung-kuo yüan yü hsi-fang ho-tso k'ai-fa shih-yu chin-shu k'uang-ts'ang" [China is willing to develop her oil, metal, and mineral resources jointly with the West], E40
"CIA Discounts China's Oil Role," C299
"CIA Doubts China to Be a Major Crude Exporter," C391
"CIA's Cautious Assessment," C393
"Closer Links with the Outside World," C488
"Coastal State Gas Signs an Agreement to Buy Chinese Oil," C540
"Coastal to Hike Chinese Oil Purchase," C601
"Combining Urban and Rural Life--Taching Impressions (IV)," C373
"Communist China and Petroleum," C56
Communist China's Industry and Materials in Translation: Petroleum, B8
"Communist China's Oil Exports: A Critical Evaluation," C237
"Communist China's Oil Exports Revisited," C334
"Communist China's Oil Industry," C21
Communist China's Petroleum Situation, B11
A Compiled Report of 1958 Petroleum Production and Refining Activities in China, B9
"A Conception of the Evolution of the Island of Taiwan and its Bearing on the Development of the Neogene Sedimentary Basins on its Western Side," C96
"Conoco Hits Gas and Condensate off Taiwan," C202
"Council Visits Taching," C345
"Country Report: China (Taiwan)," C111
"CPC (Chinese Petroleum Corporation) Goes Petrochemical," C458
"Criticism of Teng Hsiao-ping Is in a New Upsurge in Taching Oil Field," C398

"Danger Zones in the South China Sea," C426
"Development in Mainland China, 1949-1968," C99
"Development of Man-Made Fibre Industry and Petrochemical Industry in Taiwan," C159

"Development of Petrochemical Industries in Republic of China," C368
<u>Digest of Coal, Iron, and Oil in the Far East</u>, B3
"Doing Business with China," C586
<u>Doing Business with the People's Republic of China: Industries and
    Markets</u>, B47
"Drilling Pushed in China's Pohai Gulf," C174
"Dutch, U.K., U.S. Firms Vie for Petroleum Equipment Sales," C367

"Earthquake Strikes at China's Energy Centers," C346
"East Asia and Australia: A Remarkable Development: China," C427
"East Asian Coasts, Offshore Are Promising Petroleum Frontiers," C349
<u>Economic Development and Use of Energy Resources in Communist China</u>,
    B12
<u>An Economic Geography of China</u>, B14
"Energy and Power" (in <u>China Trade Report</u>, April 1979), C594
"Energy and Power" (in <u>China Trade Report</u>, Aug. 1979), C622
"Energy and Power" (in <u>China Trade Report</u>, Nov. 1979), C640
"Energy in China: Achievements and Prospects," C311
"Energy in China: Interview with Industry Officials," C599
"Energy in the People's Republic of China," C188
"Energy in the P.R.C.," C306
<u>Energy Policies of the World</u>, B34
"Energy Resources of China," C379
"The Energy Situation in Taiwan, Republic of China--Past, Present,
    and Future," C216
"Energy Solution in China," C404
"The Era When We Depend on 'Foreign Oil' Is Gone for Good," C32
Erh-shih nien lai chih Chung-kuo shih-yu [Chinese petroleum
    corporation in the last twenty-five years], D24
"Esso: Far East Oil Potential Limited," C481
"Excellent Situation in Taching Oilfield," C72
"Explorations: Another Player in the Oil Game," C212
"Explorations in China," C1
"Exxon Unit Contracts with China to Conduct Offshore Examination,"
    C605

"Fair Prospects," C448
"Far East--The Focus Is on China," C624
"Fast Boat to China (Polypropylene Resin)," C157
"Feasible New Petrochemical Projects in the ROC," C445
"15,000-Ton Oil Tanker Launched in China," C84
"Fill'er Up," C51
"Financial Snags Are Seen Allowing China in Buying Japan's Steel,
    Oil Technology," C587
"Firms Poised for China Seismic Surveys," C612
"First Geological Prospecting Ship--New Victories on the Oil Front,"
    C243
"First in China Trade (Acrylonitrile)," C166
"5,000-Ton Tanker Built on Sandy Bank," C123
"Foraminiferal Trends in the Surface Sediments of Taiwan Strait,"
    C114

"For More than Oil--Taching Impressions (I)," C369

"4 U.S. Oil Companies, with Official Support, Start Talks with China," C470

"From Dependence on 'Foreign Oil' to Basic Self-Sufficiency in Petroleum," C35

"Fruit of Great Cultural Revolution: Taching Is Five Times Its Former Self," C192

"Fuel Industry: Petroleum," C140

"Fukyō to Chūgoku to sekiyu sangyō (Shinshun kandan)" [Economic slump, China, and the petroleum industry (New Year's Discussion)], E24

"Further Oil Strides by China Reported in Trade Magazine," C645

"Gas for Fuel and Industry," C118

"Gas for Industry and Home Use," C196

Gendai no Chūgoku keizai: sekiyu to shakaishugi kensetsu [Contemporary Chinese economics: petroleum and the building of socialism], D38

General Information of Chinese Petroleum Corporation, B5

"Genyu no tai-Nichi yushutsu gonengo sen-gohyaku man ton" [Crude oil exports to Japan will reach fifteen million tons in five years], E37

"Genzoic Basin of Western Taiwan--Case History of the Chingtsaohu Gas Field, Taiwan, China," C107

"Geological Structure and Some Water Characteristics of the East China Sea and the Yellow Sea," C82

"Geologic Concepts Relating to the Petroleum Prospects of Taiwan Strait," C66

"The Geologic of China's Oil," C591

"Geology of Gu-dao Oil Field and Adjacent Areas," C589

"Geophysics in China," C546

"The Geopolitics of China's Oil," C443

"George Bush: Oil for China's Self-Reliance," C436

"Geosource Plans to Sell Oil Equipment to China," C512

"German's Uhde Seen Winner of Key Chinese Contracts," C466

"Getting Oil at High Speed--Notes on Taching (1)," C395

"Giant Gas Tank," C215

"Great Victories for China's Petroleum Industry," C105

"Group of Japan Firms Will Export to China 4 Petrochemical Plants," C563

"Growth Seen in China's Oil Equipment Buying," C343

"Guided by Mao Tse-tung's Thought, China Introduces Fermentation Dewaxing Process in Oil," C53

"Hakusha kakaru Chūgoku no sekiyu kaihatsu" [China's oil development in high gear], E14

"Has China Enough Oil?," C36

"Hercules to Taiwan," C152

"High Output Oil Field Developed in China," C629

"Highway-Pipeline Double-Utility Bridge," C552

"'Historic territorial rights': Key to China's Oil Future," C307

"Hong Kong," C506
"Ho-pei Ts'ang-chou ch'eng shih-yu ch'eng" [Ts'angchou in Hopei
    province has become an oil city], E50
"The Houston-Peking Axis: Oil for China's Lamps--and More," C296
"How China Developed her Oil Industry," C205
"How China's First Long Petroleum Pipeline Was Built," C232
"How Much Oil?," C45
"How Rolligon Sold Petroleum Equipment to China," C410
"Hsi-k'an Chung-Jih shih-yu kuan-hsi" [A close look at the Sino-
    Japanese oil relationship], E39
"Hsin-hsin-hsiang-jung-ti Ta-ch'ing yu-t'ien" [The prospering
    Tach'ing oil field], E38

"ICI Gets Invitation to Participate in Taiwan in Plastic Venture,"
    C501
"Ideology and Oil," C229
Implications of Prospective Chinese Petroleum Development to 1980,
    B35
Impressions of Taching Oil Field, B48
"Improved Outlook for Chinese Oil Output Dollar Earnings, Buying,"
    C384
"India Considers China as Source of Crude," C643
"Industrial Waste Water Serves Agriculture," C134
"Industry Profile: Oil," C274
"In Pursuit of Oil," C625
"Inside Look at China's HPI," C189
"Internal Strife May Hurt Communist Chinese Oil Effort," C57
The International Energy Policies of the People's Republic of China,
    B36
The International Energy Relations of China, B53
"International Report: People's Republic of China," C595
Investment in Taiwan: A Look at the Petrochemical Industry and
    United States Involvement in its Growth and Development, B15
"Isocyanates Unit Bought by China from Japan," C492
"Iwayuru 'Chūgoku sekiyu' no tembō"[The prospects for what is called
    'China's oil'], E30

"Japan and China Will Take Up Development of Oil," C476
"Japan, China Plan Offshore Venture," C473
"Japan-China Trade: Oil and Euphoria," C505
"Japan-China Trade Up 40% with Plastic Resins Leading the Way," C465
"Japanese Firms Face Tough Problems in Using Chinese Oil, Coal," C530
"Japan, Final Terms for Chinese Oil," C437
"Japan: Mixing Oil and Steel," C364
"Japan Peace Treaty's Effect on the Oil Industry," C509
"Japan Refinery Seeks Big Funds--Has Eye on Chinese Oil," C568
"Japan Seeks Longer-Term Oil," C284
"Japan Sends Oil Mission to China," C401
"Jenchin (Jenchiu)--A New Big Oilfield in China," C485
"Jenchiu--A New High-Yielding Oilfield," C516

"Jen-ch'iu yu-t'ien chien ch'eng" [Building of the Jench'iu oil field is completed], E48
"JGC/Marubeni Win Chinese Ethylene Plant Contract," C467
"'Jiryoku kōsei' no riron--Taikei yuden no shisō kōzō (Chūgoku repōto" [The theory of 'self-reliance'--ideological framwork of Taching oil field (report on China)], E28

"Kanematsu-Gosho Purchases Light Oil from China," C536
Kanto-shō Ōseiken Rashikō yuketsugan chōsa hōkoku [Research report on oil shales from Kanto-shō Ōseiken Rashikō], D9
Kao-chü Mao chu-hsi-ti wei-ta ch'i chih tsou wo kuo tzu-chi kung-yeh fa-chan-ti tao-lu: Ta-ch'ing yu-t'ien tang-wei shu-chi ke-wei-hui chu-jen Sung Chen-ming tsai ch'üan-kuo kung-yeh Ta-ch'ing hui-i shang-ti fa-yen [Raising the great banner of Chairman Mao and walking our own road to industrial development: a speech given by Sung Chen-ming, Director of the Revolutionary Committee and Party Secretary at Tach'ing oil field at the Conference on State Industries held at Tach'ing], D41
"Karamai--China's New Oil Town," C17
"Key Projects Under Construction," C593
"Kono me de mita Taikei yuden--Chūgoku kokudo kaihatsu no ichi moderu" (An eyewitness report on the Taching oil field--China's developmental model], E13
Kung-fei shih-yu kung-yeh yen-chiu [An inquiry into the Chinese Communist oil industry], D12

"Lamps of China to get more oil, but not U.S.," C577
"Lan-chou Oil Refinery: An Example of Chinese Road to Industrialization," C47
"Large Modern Petrochemical Base," C365
"Large Oil Terminal Is Opened in China," C321
"Larger Exports Envisaged," C238
"Last Untapped Pools: Late-Comer China Emerging as Giant Petroleum Power," C539
"Learning from Ta-ch'ing: China's Oil Prospects," C399
"Legs and Walking Stick--A Visit to the Qianjin Chemical Work," C619
"Life at an Oil Refinery near Peking," C183
"Limits on China's Oil Exports," C355
"Liquid Assets: China's City of Taching Abounds with Ducks, Hogs--and also Oil," C260
"Li Szu-kuang; Prominent Chinese Geologist Dies," C115
"Look beyond the North Sea and Sell Hardware to China," C259
"Lu-Ning Shu-yu-kuan-tao chien ch'eng" [Building of the Shantung-Nanking oil pipe line completed], E46
"The Lure of Chinese Oil," C650
"Lure of the Offshore," C283
"Lurgi Gets Contract Valued at $540 Million for 5 Plants in China," C572

"The Oil Question," C184
"Oil: Statistics and Projections," C122
"Oil Up, Steel Down?," C251
"An Old Szechwan Asset for New Markets," C449
"On the Feature of Turbidite Sequences in Some Religions of China,"
    C415
"On the Geology of the Cenozoic Geosyncline in Middle and Northern
    Taiwan (China) and Its Petroleum Potentialities," C40
"On the Taching Oilfield," C164

The People's Republic of China: A New Industrial Power with a Strong
    Mineral Base, B27
"Perspectives on Energy in the People's Republic of China," C352
"The Petrochemical Industry in Taiwan--Overall Planning," C194
"The Petrochemical Industry of Taiwan--Raw Materials," C156
"Petrochemicals," C176
"Petrochemicals Getting Top Taiwan Priority," C103
"Petrochemical Works in Peking and Shanghai," C292
"A Petrochemical Works under Construction," C135
"Petroleum and Coal-Mining Industries' Achievements," C354
"Petroleum and Electric Power Industries," C26
Petroleum and Gas Deposits in the Chinese People's Republic, B7
"Petroleum--China," C3
"Petroleum Developments in Far East in 1978," C633
"Petroleum Geology and Industry of the People's Republic of China,"
    C348
"Petroleum in China," C37
"Petroleum in Kansu Province, China," C8
"Petroleum in Taiwan, B6
"Petroleum Industry--China's Economy at a Glance," C153   t of
    Energy Resources: A Bibliography with Selective Annotations, A31
"Petroleum Industry in Taiwan," C22
"Petroleum Industry in the Chinese Mainland," C38
"Petroleum Industry Keeps Advancing," C406
"Petroleum Industry, 1958-1961," C27
The Petroleum Industry of the People's Republic of China, B28
"Petroleum Industry Success-Progress Report," C124
"Petroleum Output Increases," C316
"Petroleum Power of China," C30
"Petroleum Output Increases," C316
"Petroleum Power of China," C30
"Petroleum Resources and Production in Mainland China," C31
"Petroleum Resources: How Much Oil and Where?," C241
"Photogeological Observations on the Low Hilly Terrain and Coastal
    Plain Area of Hsinchu, Taiwan," C62
"Pipeline for Shipping Crude from Shengli Oil Fields to Nanking
    Harbor Completed," C527
"Pipeline from the Taching Oil Field--New Victories on the Oil
    Front (I)," C245
"Po-hai Chu-chiang ts'ai yu chi-hua Chung-Jih ta-ch'eng ho-tso
    hsieh-yi" [Sino-Japanese accord on planning oil exploration off
    Pohai and Pearl River], E45

"Po-hai-wan yu-t'ien nien ch'an-liang mu-piao wei san-ch'ien-wan
    kung tun" [Target of annual oil production at Pohai Gulf oil
    fields set at 30 million tons], E44
"Pohai No. 1 Offshore Drilling Rig," C249
"Political Implications of the Petroleum Industry in China," C319
"Politics of China's Oil Weapon--China: The Next Oil Giant," C279
"Polyester Facilities Mulled in China," C502
"Potential Giant: Peking Experts Visit U.S.," C421
P'ou-shih kung-fei shih-yu lien chih chi-shu yü ch'an-p'in p'in-chih
    [The analysis of petroleum refining technology and products
    quality in China], D30
"Pouring Trouble on Oily Waters," C628
"Practice First: A Chinese News Report," C185
"The PRC as an Oil Exporter," C308
PRC Oil: For the Lamps of China?, B19
"Preliminary Study of Submarine Geology of China's East Sea and the
    Southern Yellow Sea," C93
"Production at Takang Oilfield Returns to Normal after Earthquake,"
    C332
"Promise of Major Finds in China's Coastal Area Draws U.S. Oil Firms;
    Asian Bonanza?," C519
"Pullman Division Gets Chinese Job for Work on Petrochemical Unit,"
    C561
"Pullman Unit Set to Aid Major Chinese Oil Project," C562

"The Quantity of Oil and the Quality of Life," C255

"Rapid Expansion of Petroleum Industry," C168
"Rapid Progress in Marine Geological Survey," C322
"Recent Development in the Petroleum Industry," C34
"Record Oil Output," C392
"Red China Battles for Oil Status," C58
"Red China Claims Large Oil Resources Being Developed," C25
"Red China Claims Oil Self-sufficiency," C86
"Red China Near Self-sufficient in Oil," C101
"Red China Reports Higher Production," C79
"Red China Reports Taching Output Up," C92
"Red China's 'Lost' Oil Field Is Found," C77
"Relation of the Tectonics of Eastern China to the India-Eurasia
    Collision: Application of Slip-line Field Theory to Large-Scale
    Continental Tectonics," C366
"Reported Oil Accord of China, U.S. Firms Is Called Premature," C497
Report on Geological Investigation of Some Oil Fields in Sinkiang, B4
"Reports on the Seismic Refraction Survey on Land in the Western
    Part of Taiwan, Republic of China," C83
Research, Production, and Utilization of Petroleum in Communist
    China, B10
"Resolutely Follow the Taching Road," C254
"Riding on China's Pendulum," C511
"The Rise of China's Oil Industry," C173

Title Index

"Russian Press Attacks U.S. Role in China's Oil Buildup," C534
"Russia Steps Up Anti-China Campaign," C600

"Saikin ni okeru Chūgoku keizai no ugoki--sekiyu kanren sangyō no
    hattan o chūshin ni" [Recent trends in China's economy--focusing
    on the development of petroleum-related industries], E12
"Sales of Chinese Oil to Power Firms," C515
"Sanyukoku Chūgoku no shōrai" [China's future as an oil producer],
    E21
"Science in China: A New Long March Begins," C517
"Scientific Experiment Promotes Development of Petroleum Production,"
    C250
"Scientific Research at Taching," C471
"The Scramble to Exploit China's Oil Reserves," C522
"Search for Oil," C642
"Search for Oil in Manchu Kuo," C5
"Second Plant Finished for Taiwan Firm," C433
"Sediments of Taiwan Strait and the Southern Part of the Taiwan
    Basin," C142
Sekiyu sangyō ni miru Chūgoku no jiryoku kōsei [China's self-reliance
    as seen in her petroleum industry], D19
"Sekiyu taikoku o mezasu Chūgoku no kaihatsu senryaku" [China's
    developmental strategy aims for leadership in oil production], E4
Sen-kyūhyaku-shichijūyonen no Chūgoku Keizai no ugoki: sekiyukanren
    sangyō o chūshin ni shite: Chūgoku .eizai kankei chōsa hōkoku
    sho [The trend of China's economy in 1974, as focused on the
    petroleum-related industries: a research report on the Chinese
    economy], D36
Sensei-shō yuden chōsa shiryō [Research data on oil fields in Shensi
    province], D1
"Shanghai Builds General Petrochemical Works--New Constructions,"
    C270
"The Shanghai No. 1 Petroleum Machinery Plant," C422
"The Shanghai Petrochemical Complex," C553
"Shangtung--Nanking Oil Pipeline Laid," C486
"Shengli, China's Second Large Oilfield," C219
"Shengli Crude is Key Component in Warm Peking-Manila Relations,"
    C441
"Shengli Journal," C423
"The Shengli Oil Field," C429
"Shengli Oilfield Thrives--New Achievement in Socialist
    Construction," C210
"Sheng-li yu-t'ien tseng-ch'an hsün-su [Production at the Shengli
    oil field rises rapidly], E42
Shih-yu jen shih hua [Discourses on the history of the Chinese
    Petroleum Corporation by its executives], D25
Shih-yu t'an-k'an [Petroleum exploration], D29
Shih-yu ti-chih hsüeh lun wen-chi [A collection of papers on
    petroleum geology], D18
Shina yuden chōsa shiryō [Research data on China's oil fields], D6

Title Index

"Shogaikoku kara mita Chūgoku no sekiyu jijō" [China's oil situation
    as viewed by various foreign nations], E17
"Shōri yuden--daini no Taikei" [Shengli oil field--the second
    Taching], E47
Shōwa jūninendo Sansei yuboketsugan oyobi sekitanchi shisui
    chishitsu chōsa hōkoku [Research report on the trial drilling of
    coal and oil shales at Sansei in 1937], D7
"Sichuan Journal--Natural Gas," C647
"Sino Accent: Conservation, Exploration," C654
"A Sino-Japanese Crude Oil Contract, 1974," C313
"The Sino-Japanese-Korean Sea Resources Controversy and the
    Hypothesis of a 200-Mile Economic Zone," C294
"Sino-Japanese Oil Drilling Will Be Tried in Pohai Gulf," C538
"Sino-Soviet Relations and Politics of Oil," C320
"A 6,011-Meter Deep Well," C317
"Slowing Down or Dying--The Oil Boom--China '76," C340
"Snags Facing China's Oil Exports," C288
"Sobering Thought in the Oil Rush," C531
"Solving the Chinese Oil Puzzle," C450
"South China Sea Oil: Opportunities Amidst Political Uncertainties,"
    C637
"South China Sea Oil Search Mixes Economics, Politics," C626
South China Sea Oil: Two Problems of Ownership and Development, B44
"Southern Asia: China Prepares for Massive Oil and Gas Development,"
    C528
"Soviet, Chinese Drilling Shows Different Emphasis," C513
"Soviets Doubt Accuracy of Chinese Oil Production Figures," C655
"Soviets Hint Red China's Taching is a Paper Tiger," C59
"Soviets Hit China Oil Potential Reports," C580
"Soviet Union Refuses to Concede China Has Huge Offshore Reserves,"
    C201
"Split with China Jolts Albanian Oil Industry," C479
"Stand Up Straight--Taching Impressions (II)," C371
"The Story of Iron Man Wang," C396
"A Stratigraphic and Sedimentary Analysis of the Protoquartzite in
    the Miocene Talu Shale in Northern Taiwan," C63
"Striving for Ten New Tachings," C461
"Structural Framework of East China Sea and Yellow Sea," C100
Studies on Tectonics and Petroleum in the Yantse (Yangtse) Region of
    Tshung-king (Chung-king), B1
"A Study of the Energy Sources on the China Mainland," C246
"A Study of the Geology and Petroleum Potentialities of Paoshan and
    Chingtsaohu Area, Hsinchu," C29
"Successes Spur Chinese Search Onshore and Off," C646
The Survey of the Energy Economy of Taiwan, B13
"A Swell of Oil Disputes in the South China Sea," C635
"Szechuan Gas Fields--New Victories on the Oil Front (II)," C247
"Szechwan Province Speeds Up Development of Gas Fields," C271
"Szechuan's Gas Fields Making Headway," C253

"Tachai and Taching," C424
Ta-ch'ing [Tach'ing], D32
"Taching Chemical Fertilizer Plant," C356
"Taching Chemical Fertilizer Plant Goes into Operation," C409
"Taching--China's Model for Industry," C457
"Taching Fights the 'Four Pests,'" C382
"Ta-ch'ing jeng k'o kai-ts'ai san-shih nien" [Tach'ing oil field can
    still last another thirty years], E43
"Taching No. 30 Oil Tanker Built," C127
"The Taching Oil Field," C108
"Taching Oil Field Develops under the Direction of the Thought of
    Mao Tse-tung," C44
The Taching Oilfield: A Maoist Model for Economic Development, B20
"Taching Oilfield, Pt. I: Race against Time," C110
"Taching Oilfield, Pt. II: The Women," C117
"Taching Oilfield, Pt. III: Death of an Oligopoly," C119
"Taching Oilfield's Fresh Victories," C146
"Taching Oilfield Wins Tremendous Success in Revolution and
    Production," C91
"Taching--Oilfield with Both Rural and Urban Characteristics," C377
"Taching Oil Hikes Red Chinese Output," C81
"Taching Overfulfills Quarterly Production Plan," C376
"Taching Red Banner Becomes More Radiant Than Ever," C113
"Taching--Red Banner on China's Industrial Front" (in Economic
    Reporter [English supplement]), C378
Taching: Red Banner on China's Industrial Front (Foreign Language
    Press [Peking]), B17
"Taching's Red Banner Flies Even Higher," C235
"Taching's Theme Is Growth," C370
"Taching Today," C381
"Taching Workers Courageously Stride Forward," C78
Ta-ch'ing yu-t'ien chi-shu ke-hsin tzu-liao hsüan [Selected materials
    on technological innovations at Tach'ing oil field], D45
"Taikei no akahata" [Taching's red banner], E31
"Taikei yuden no mezasu mono--kōshin kara senshin e no migoto na
    tenshin" [The aim of the Taching oil field--an excellent
    transformation from the underdeveloped to the advanced], E16
"Taikō yuden no kembun" [Observations from the Takang oil field], E19
"Taiwan and Korea Load Up for Exports," C451
"Taiwan Chemicals Trade Boom as Export Sales Hit $8.2 Billion," C387
"Taiwan Ethylene Plant On-Stream," C73
"Taiwan Industries Feeling Oil Crunch," C610
"Taiwan no sekiyu kagaku kōgyō [Petrolchemical industry in Taiwan],
    E29
"Taiwan Plans $103.6 Million Oil Program," C116
T'ai-wan shih-yu [Petroleum in Taiwan], D15
T'ai-wan shih-yu chi t'ien-jan-ch'i chih t'an-k'an yü k'ai-fa [The
    exploration for and development of petroleum and natural gas in
    Taiwan], D26
T'ai-wan shih-yu hua-hsüeh kung-yeh chih chien-chieh [A summary of
    the petrochemical industry in Taiwan], D27

"T'ai-wan shih-yu t'an-k'an chi-yao" [A summary of petroleum exploration in Taiwan], D28

T'ai-wan shih-yu ti-chih t'ao-lun-hui lun-wen chuan chi [Symposium on the petroleum geology of Taiwan], D14

"Taiwan's Big Push: Petrochemicals," C69

"Taiwan's First Styrene Plant Ordered," C195

"Taiwan's Gas Hopes Evaporate," C464

"Taiwan's Petrochemical Industry in a World Context," C172

"Taiwan's Quest of Oil," C19

Taiwan yuden chōsa hōkoku [A report on the investigation of Taiwan's oil fields], D2

Ta-kang yu-t'ien chih kai-shu [A synopsis of Takang oil field], D35

Ta-lu fei-ch'u shih-yu hua-hsüeh kung-yeh [Petrochemical industry in mainland China], D34

Ta-lu fei-ch'ü shih-yu shu ch'u hsi-t'ung chih yen-chiu [A study of mainland China's petroleum transportation and storage system], D40

Tao-lu shih-yu hua-hsüeh kung-yeh [Mainland China's petrochemical industry], D43

"Ta-lu shih-yu kung-yeh chih fa-chan ch'ing-k'uang" [The condition of mainland China's petroleum industry], E33

Ta-lu shih-yu kung-yeh hsien-shih [The current condition of the petroleum industry in mainland China], D21

"Ta-lu shih-tiyu shu-chu ch'ien-li" [China's petroleum export potential], E34

Ta-lu shih-yu tzu-yüan chih-fen pu [The distribution of petroleum resources in mainland China], D31

"Tap Potentials, Increase Production," C150

"Tapping Natural Gas in Szechuan," C170

"Technology in China," C50

"Temporary Stagnation in Output," C335

"Ten Major Contributions of Taching Oilfield," C361

"Ten U.S. Companies Join Search for Oil Off China," C623

"Te o musubu Chūgoku to mejā no shiwaku" [China's connection with oil majors—a puzzle], E52

"300,000-Ton Ethylene Project," C358

"Three Japanese Groups Get Jobs from China Totaling $700 Million," C570

"A Thriving Harbour—Chinhuangtao," C309

T'ien-jan-ch'i ch'ien-tsai shih-ch'ang hsü-yao tiao-ch'a pao-kao [The potential market demand for natural gas: an investigative report], D20

Ti i ko shih-yu chi-ti—Yü-men [The first oil base—Yümen], D13

"Time Bomb in East Asia—China: The Next Oil Giant," C280

"To Develop an Oil Field, Get Rid of Ghosts and Ogres," C52

"Tokyo, Peking to Explore Oil Offshore China," C474

"Toward Worldwide Oil Sales," C548

"Trading U.S. Technology for Chinese Oil," C419

"Tsai lun Chung-kuo-ti shih-yu wen-t'i" [Another discussion of China's petroleum problems], E36

"Tui Chung-kung shih-yu sheng-ch'an shu-ch'u chi yün-shu neng-li-ti yen hsi" [An examination of Chinese Communist oil production, exports, and transportation capabilities], E22
"Turmoil Cripples Red Chinese Oil," C65
"12 Glorious Years of Taching Oilfield," C162
"1205 Team Drills over 127,000 Metres," C132
"Two Chemical Installations Go into Operation in China," C344
"Two French Firms to Build Chinese Petrochemical Works," C169
"Two New Oil Fields and a New City," C248

"Underground Oil Tank," C390
"U.S. Catches China's Eye," C447
"U.S., Chinese Meet to Swap Oil, Gas Reserves Ideas," C613
"The U.S. Discovers China," C574
"U.S. Is Said to Bar Drilling Off China," C275
"U.S. Oil Firms May Get Chance to Help China Exploit Its Offshore Deposits," C472
"U.S. Oil Concerns Step Up Drive to Prod China into Reaching Accord on Drilling," C575
"U.S. Suppliers Strengthen Their Share of China Oil Rig Market," C452
"U.S. to Get Low-Sulfur Chinese Oil--Coastal States Signs Accord for Imports," C542
"U.S. Urged to Oppose Oil Claim by Taiwan," C402
"An Up and Coming Oil Refinery," C418

"A Vigorous Force on the Petroleum Front," C252
"Vigorous Political and Ideological Work--Report from Taching Oilfield," C130
"Visit to an Oil City," C298
"A Visit to Takang Oilfield," C207
"A Visit to the Nanking Petrochemical Works," C293

"Waga michi o yuku Chūgoku sekiyu kaihatsu" [China travels her own road in oil development], E5
"War's End Brings Accelerated Activity in Yumen, China's Only Oil Field," C10
"Who Is Making the Big Business Deals with China?," C535
"Will There be Plenty of Oil after All?," C518
Winds of Change: Evolving Relations and Interests in Southeast Asia, B38
"A Women's Oil Team--Report from Taching Oilfield," C143
"Worker-Peasant Villages--Notes on Taching (2)," C397
"Worldwide Drilling and Production: China Abounds in Contracts, Projects," C611
"Worldwide Report," C350

"Xylene Rounds Out Taiwan Chemical Group," C503

"Yenchang Oil Field," C12
Yüeh chin chung ti yu-t'ien chien-she [Oil field construction is advancing], D33

"Yumen Oil Field Plays Its Part," C603
"Yung Mao chu-hsi che-hsüeh ssu-hsiang chih-tao chao yu chi"
    [Chairman Mao's philosophical thought guides us in search of
    oil and gas], E9